Matheus Henrique Glória de F. Cardoso

Development of software for generating and adjusting TTT diagrams

Matheus Henrique Glória de F. Cardoso

Development of software for generating and adjusting TTT diagrams

Isothermal transformation of hypoeutectoid steels

ScienciaScripts

Imprint
Any brand names and product names mentioned in this book are subject to trademark, brand or patent protection and are trademarks or registered trademarks of their respective holders. The use of brand names, product names, common names, trade names, product descriptions etc. even without a particular marking in this work is in no way to be construed to mean that such names may be regarded as unrestricted in respect of trademark and brand protection legislation and could thus be used by anyone.

Cover image: www.ingimage.com

This book is a translation from the original published under ISBN 978-613-9-68418-2.

Publisher:
Sciencia Scripts
is a trademark of
Dodo Books Indian Ocean Ltd. and OmniScriptum S.R.L publishing group

120 High Road, East Finchley, London, N2 9ED, United Kingdom
Str. Armeneasca 28/1, office 1, Chisinau MD-2012, Republic of Moldova, Europe
Printed at: see last page
ISBN: 978-620-8-21786-0

SUMMARY

ACKNOWLEDGMENTS 2
SUMMARY 3
1 INTRODUCTION 4
2 THEORETICAL BACKGROUND 8
3 literature review 23
4 METHODOLOGY 29
5 Results and discussion 37
6 Conclusion 57
BIBLIOGRAPHICAL REFERENCES 59

ACKNOWLEDGMENTS

To God for my life, family and friends.

To my parents and sister, for their support and encouragement throughout.

To my fiancée, Fernanda, for all her affection, love and companionship.

To Professor Dr.-Ing. Pedro Paiva Brito, for his guidance in this work.

To Professors André Fioravante de Oliveira and Tarcisio Flâvio Umbelino Rego for their constant support and encouragement.

SUMMARY

One of the main factors influencing the mechanical behavior of steel components is the rate of cooling from the austenitic field, which can lead to the formation of different microconstituents. The type of microstructure formed in a steel as a function of the cooling rate can be known, in turn, by analyzing isothermal transformation diagrams (or time-temperature-transformation diagrams, TTT) which show the time required for the decomposition of austenite to occur at a given temperature. These diagrams are characteristic of a given steel and depend to a large extent on its chemical composition. Due to the complexity involved in the experimental determination of TTT diagrams and the continuous development of new high-strength hardenable steels, since the 1970s the development of mathematical models has been sought that allow the determination of theoretical TTT diagrams based on the chemical composition of a steel. In this study, *software* was developed to model the TTT diagram and, based on this, determine the Jominy hardenability curve of a steel with a known chemical composition. The *software* developed was a platform that enabled the implementation of a numerical tool to correct the modeled TTT diagram by comparing the theoretically calculated hardenability curve with experimental data. Using the methodology proposed in this work, it was possible to obtain corrected TTT diagrams, with a reduction in the mean absolute deviation in relation to the initial model, from the laboratory hardenability test for ABNT 1045, 4140 and 4340 steels.

Keywords: Steel, Jominy hardenability, TTT diagram, microstructure modeling.

1 INTRODUCTION

Steels are iron-based alloys that are currently used extensively in various industries to manufacture a wide range of products. One of the reasons for the widespread use of steels is their versatility, which is expressed in the wide range of physical properties that can be obtained depending on their chemical composition and the processing conditions to which they are subjected. It should also be noted that iron is an abundant element in the earth's crust, has a low production cost for general applications and has undergone continuous evolution in recent years in the synthesis of the metal alloys it forms, allowing for greater control of chemical composition and microstructure.

There is currently a strong trend towards reducing the weight of metal structures and mechanical components, especially in the automotive and aerospace industries. In machine elements made of steel, the desired weight reduction can be achieved by increasing the mechanical strength of the material used. Traditionally, to achieve this increase in strength in carbon or alloy steels, quenching heat treatments are used, whether or not followed by tempering.

The purpose of quenching a steel is to increase its hardness by forming martensite and consists of heating the steel to a suitable temperature, which allows the austenitic microstructure to be established homogeneously, followed by sudden cooling, usually in an oil or water bath (CALLISTER, 2007, p. 331). Martensite has a high hardness (depending on the carbon content of the steel) and high cooling rates are required for its formation. In fact, a slowly cooled steel, as in annealing or normalizing treatments, has a microstructure made up of ferrite and pearlite and can have a hardness 2 or 3 times lower than in a tempered condition (CHIAVERINI, 2008). Therefore, in order for there to be homogeneous hardening of a hardened component, there must be a high enough cooling rate throughout its entire volume for martensite to form. In this sense, controlling phase transformations in the solid state, the nature of which depends on the cooling rate at which they occur, is a tool for adapting the microstructure of steels and thus adjusting their mechanical properties to obtain a wide range of strength and ductility values (PERELOMA and EDMONDS, 2012).

On the surfaces of a hardened component, the high rates required for martensite formation are normally achieved through contact with the cooling medium. However, in larger cross-section parts, the central regions are cooled more slowly than those located on the surface, thus preventing uniform hardening throughout the volume of the component. The ability of a steel to increase its hardness evenly when cooled rapidly is called hardenability and this characteristic must be taken into account when selecting a steel for hardening. One of the most common ways of assessing the hardenability of a steel is through the Jominy test, which consists of taking a previously austenitized cylindrical specimen and cooling it with a jet of water falling on its base. The hardenability assessment is made by drawing the Jominy curve, a diagram that relates the hardness values of the cylinder on the vertical axis and the position of the hardness in relation to the end that received the jet of water on the horizontal axis.

In order to increase the hardenability of steel, allowing for the tempering of larger pieces, alloying elements are added to the material, such as: Mo, Nb, Mn, Ni, Cr, etc. These alloying elements delay the decomposition of austenite, allowing martensite to form at slower cooling rates (COLPAERT, 2008, p. 282). Important information on the microstructure as a function of the cooling rate of a steel can be obtained from isothermal transformation (IT) diagrams, also known as time-temperature transformation (TTT) diagrams. These diagrams are obtained experimentally in a laborious way and are not readily available for new steels, or even for existing steels, but are of specific use (COLPAERT, 2008, p. 202).

Due to the importance of chemical composition in the hardenability of steels, since the 1970s, researchers have sought to develop models that allow theoretical TTT diagrams to be drawn up. A pioneer in this field was Kirkaldy (1978), who developed a computational method for obtaining the TTT diagram. In his approach, he sought to correlate the temperatures Ae3 (the cooling temperature that marks the start of austenite decomposition) for steels with Mn, Si, Cr, Mo and Cu alloying elements and the times required (as a function of temperature) for austenite decomposition to be completed. Experimental modeling of hardenability has also been attempted, for example in the work by Andrés and Carsi (1987), in which an experimental model was proposed for the Jominy curve based on the steel's chemical composition and austenitic grain size. As studies progressed, however, the methodology adopted by Kirkaldy was maintained, i.e. obtaining the TTT diagrams from the austenite decomposition kinetics equations, as evidenced by the subsequent refinements of the original approach proposed by Umemoto *et al.* (1980) Lee and Bhadeshia (1993) and Victor Li *et al.*

(1998). The latter, Victor Li *et al.* (1998), developed a model for the TTT diagram based on the chemical composition of alloy steels, which was integrated with a heat transfer model from the Jominy test, allowing Jominy curves to be drawn from the superimposition of cooling rates on the TTT diagrams and the fractions of constituents formed for each cooling rate to be calculated. Three equations were used to model the kinetic reaction of austenite decomposition, for ferrite, pearlite and bainite respectively. More recently, Zhanli Guo *et al.* (2009), following the same conceptual basis, developed a model for phase transformations and material properties that are critical for predicting distortions in the heat treatment of steels. Unlike Kirkaldy, who was successful only with low-alloy steels, this new model, according to the authors, can be applied to a wider range of steels. Other recent efforts include work by Trzaska and Dobrzanski (2007), Trzaska and Dobrzanski (2013), Sushanthi and Maity (2014) and Bok *et al.* (2015).

A clear trend that has emerged from these studies is that the final properties of cooled steels are obtained from the fractions and hardnesses of the constituents (ferrite, pearlite, bainite and martensite) present in the microstructure. In turn, the fractions of the constituents are obtained by integrating the transformed fraction predicted by the decomposition kinetics equations over time, thus allowing virtual Jominy curves to be obtained (UMEMOTO *et al.*, 1980; VICTOR LI *et al.*, 1998; ZHANLI and GUO, 2009). Comparison with experimentally obtained hardenability curves could then be used to assess the general model used to determine the TTT diagram. However, there has not yet been a proposal to take the opposite route, i.e. to correct the virtual TTT diagrams from the experimentally obtained hardenability curves. In this way, the aim of this study was to develop a computational tool that would allow virtual TTT diagrams to be adjusted based on the hardenability test (Jominy) .[1]

1.1 General Objective

Develop a computational tool for modeling the TTT diagram of hypoeutectoid steels, allowing its indirect correction by comparing virtual Jominy hardenability curves, generated from the diagram to be corrected, with experimental hardenability data.

1.2 Specific Objectives

- Implement a computer routine to generate the TTT diagram of a steel based on its chemical composition;

- Implement a computer routine to generate a virtual Jominy hardenability curve from the modeled TTT diagram;

- Develop an algorithm for adjusting the virtual Jominy hardenability curve by refining the calculation parameters of the TTT curve model;

- Validate the effectiveness of the algorithm developed by comparing the corrected TTT diagrams with known diagrams;

1.3 Justification

TTT diagrams are important tools for predicting the microstructure obtained in steel components subjected to tempering heat treatment. However, in addition to these diagrams being unavailable for certain steels, they are sensitive to variations in chemical composition, austenitic grain size, part thickness and tempering severity, so that there may be discrepancies between the reference information in the literature and the expected behavior of a given steel. By applying the methodology proposed in this work, it will be possible to develop a computational tool which, based on a simple hardenability test, can generate a realistic estimate of the TTT diagram of a given steel. In addition

1 TTT diagrams only predict transformations that occur at constant temperature, which is not observed during the Jominy test. The details of how this is dealt with in this work are presented in Chapter 3 of this work, Methodology.

to the technical-scientific merit of the work, which is justified by the unprecedented approach to be used, the possibility of making the *software* available to the academic community in general for teaching and research purposes should be highlighted.

1.4 Scope of work

This text is organized into the following chapters, apart from this Introduction:

- Theoretical background: basic concepts related to the development of the proposed topic are presented, involving elements of phase transformation kinetics, presentation of TTT diagrams and notions of steel hardenability;

- Bibliographical review: this presents the state of the art on the subject developed in this work, seeking to present both the classic articles that initially drove research on the subject and more recent studies;

- Numerical methodology: detailed description of the steps required to develop the algorithms to be implemented in this work;

- Experimental methodology: presentation of the plan of experiments to be carried out to validate the computational routines implemented;

- Results and discussion: presentation and discussion of the routines already implemented;

- Conclusions: summary of the most important results obtained during the research and suggestions for future work.

2 THEORETICAL BACKGROUND

This chapter presents the concepts that form the theoretical basis of this work. The aim of this exposition is to present the physical principles related to phase transformations that occur in the solid state, including formulations for the kinetics of these reactions, TTT diagrams and the concepts of steel hardenability.

2.1 Phase transformations in the solid state

Many phase transformations that occur in materials have the atomic diffusion process as their predominant mechanism. These transformations, which include, for example, solidification, recrystallization and the decomposition of austenite under equilibrium conditions, occur in two distinct stages: nucleation, which can be homogeneous or heterogeneous, and growth. The nucleation stage consists of the formation of the first stable volumes of the phase transformation products, while the growth stage takes place, simply put, by the diffusion of atoms from the volume occupied by the original phase to the volume occupied by the new phase of the transformation (CALLISTER, 2007).

2.1.1 Homogeneous nucleation

The driving force for phase change is the reduction of the material's free energy. For transformations occurring at constant pressure, the free energy formulation is:

$$\boldsymbol{G = H - TS} \tag{1}$$

where G is the Gibbs free energy, H the enthalpy of the system, T the absolute temperature and S the entropy. Conceptually, Gibbs free energy represents the maximum amount of energy that can be extracted from a system without changing its volume or flowing heat through its boundaries. Physically, the free energy is the sum of the kinetic and potential energies of the fundamental particles that make up the system (in the case of a material, its atoms). Therefore, the change in free energy (*FGE*) between the final and initial stages of a transformation represents the amount of energy released by the material as a result of the transformation. Therefore, for a transformation to occur spontaneously in a material, $AG < 0$ is required (PORTER and EASTERLING, 1992).

For the description of the homogeneous nucleation process, consider the phase transformation expressed by Equation (2):

$$\beta \rightarrow \alpha \tag{2}$$

where β and a designate, respectively, the phases present before and after the transformation. Suppose, furthermore, that for the hypothetical system considered here, a and β are solid phases and that the transformation represented in Eq. (2) takes place on cooling to a certain equilibrium temperature $T_{.e}$

Figure 1 shows the Gibbs free energy variation curves for the *a* and *β* phases as *a* function of temperature. It can be seen that when the temperature of the material is equal to *Te*, the free energy of both phases is the same. Under this condition, the system is in equilibrium, which means that the rate of conversion of *β* into *a* is equal to the rate of conversion of *a* into *β*, i.e. the transformation expressed in Eq. (2) does not occur. In order for *a* to form predominantly, the material must therefore be at a temperature where the free energy associated with the *a* phase (G_a) is less than the free energy associated with the *β* phase (G_β), so that $\Delta G = G_a - G_\beta < 0$. The minimum variation in free energy required for the phase transformation to take place (ΔG^*) is called the activation energy for the process and, in order for it to be reached, the material must be at a lower temperature than the transformation equilibrium temperature. This temperature difference is called supercooling (ΔT).

Figura 1 - Evolution of Gibbs free energy for two phases *a and β* as *a* function of temperature during a phase transformation

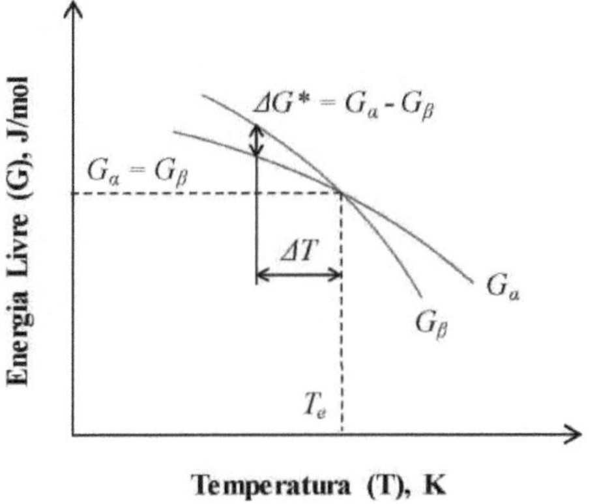

Source: Adapted from PORTER and EASTERLING (1992)

In the case of homogeneous nucleation, the first volumes of *the a-phase* of the transformation have a boundary only with the beta phase. The nuclei tend to be spherical due to the lower ratio between surface area and volume, as shown schematically in Figure 2.

Figura 2 - Representation of the process of homogeneous nucleation of phase *a* from *β*

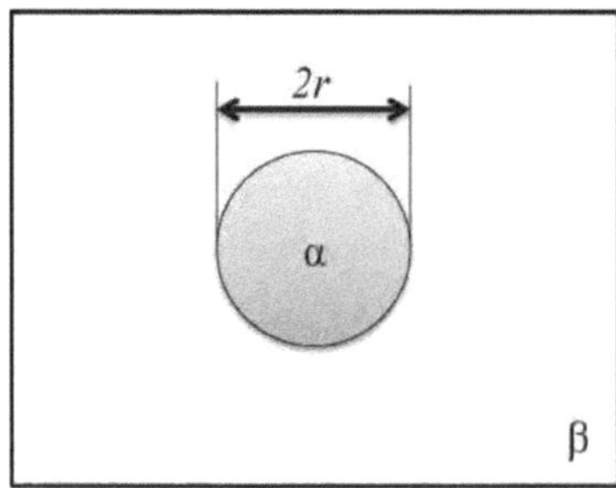

Source: Prepared by the author

In the situation shown in Figure 2, the change in free energy associated with the transformation of *β* into *a* ($\Delta G_{\alpha\beta}$) can be expressed according to Eq. (3):

$$\Delta G_{\alpha\beta} = V\Delta G_V + S\gamma_{\alpha\beta} \tag{3}$$

where V is the volume of the nucleus (m^3), ΔG_V the free energy of volume (J/m^3), S the surface area of the nucleus (m^2) and $\gamma_{\alpha\beta}$ the surface tension between phases *a* and *β* (J/m^2). The term $S\gamma_{\alpha\beta}$ represents the energy expenditure involved in creating the interface between the two phases.

The variation in the free energy of volume can be calculated by (PORTER and EASTERLING, 1992):

$$\Delta G_V = \frac{\Delta H_{\alpha\beta}\Delta T}{T_e} \tag{4}$$

where $\Delta H_{\alpha\beta}$ is the difference between the formation energies of phases *a* and *β* (negative for $T < T_e$). In Eq. (4), the contribution due to deformation energy, which is present in transformations that take place in the solid state due to the incompatibility between the crystal structures of the phases involved in the transformation, has been disregarded (PORTER and EASTERLING, 1992).

Graphically, Eq. (3) can be represented as shown in Figure 3, where the contributions of the volume free energy and the surface energy of the nucleus are related. It can be seen that there is a maximum value for the free energy variation associated with the creation of the nucleus, which represents the activation energy for the nucleation process (ΔG^*). This free energy variation value is in turn linked to a critical nucleus size of radius r^*. Both parameters (ΔG^* *and* r^*) can be obtained by

by deriving Eq. (2) as a function of *r*. For example, the activation energy is given by:

$$\Delta G^* = \frac{16\pi\gamma_{\alpha\beta}^3 T_e^2}{3\Delta H_{\alpha\beta}^2 \Delta T^2} \tag{5}$$

The result expressed in Eq. (5) shows that the activation energy for nucleation varies inversely with the square of the supercooling, which means that for transformations close to the equilibrium temperature, the activation energy is very high. Therefore, the nucleation rate is small when the nucleation temperature is similar to the transformation equilibrium temperature.

Figure 3 - Evolution of the associated free energy variation as a function of the size of the nucleus

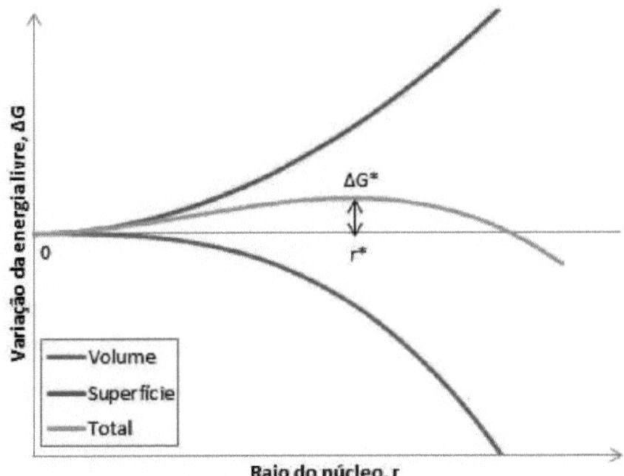

Source: Adapted from Callister (2007)

2.1.2 Heterogeneous nucleation

In heterogeneous nucleation a variable shape is obtained for the nucleus, depending on the substrate where nucleation takes place and the wettability, which is the angle of contact between the two different phases and represents the chemical affinity between them. This nucleation of the new phase usually occurs at the grain boundaries or at the junction of several grains, due to the fact that these locations provide an amount of critical energy that is needed for nucleation, thus reducing the amount of energy needed provided by the thermal variation. This is due to the reduction in the interface area between the phases involved in the transformation, since the phase transformation product is partially in contact with the heterogeneous nucleation substrate.

The equation that represents the critical energy required for heterogeneous nucleation from a planar grain boundary between two adjacent grains, as illustrated in Figure

4, is given by (CLEMM and FISHER, 1955):

$$\Delta G^* = \frac{4}{27}\frac{(b\gamma_{AB} - a\gamma_{AA})^3}{c^2 \Delta G_v^2} \quad (6)$$

where γ_{AB} is the energy at the AB interface, γ_{AA} is the energy at the AA interface and the coefficients *a, b* and *c* depend on the types of grain junctions and the calculations become more complex as the number of grains increases.

Figura 4 - New phase with symmetrical lens shape

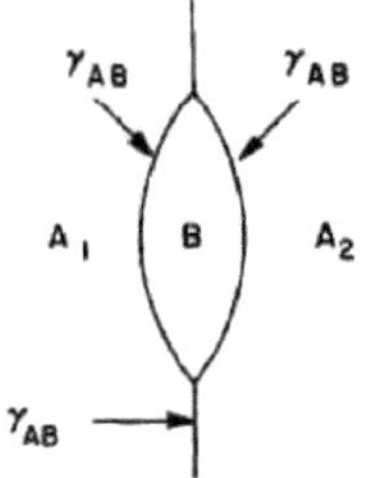

Source: CLEMM and FISHER (1955)

The parameters for this type of nucleus growth presented by Clemm and Fisher (1955) are:

$$a = \pi(1 - k^2) \tag{7}$$

$$b = 4\pi(1 - k) \tag{8}$$

$$c = \frac{2\pi}{3}(2 - 3k + k^3) \tag{9}$$

where $k = \cos\theta$ e θ is the angle of wettability.

At the junction of three grains we can see the following shape as shown in Figure 5.

Figura 5 - Volume of the new phase formed at the três grâos meeting

Source: CLEMM and FISHER (1955)

The coefficients for area and volume presented for three-gear joints are:

$$a = 3\beta(1 - k^2) - k(3 - 4k^2)^{\frac{1}{2}} \tag{10}$$

$$b = 12(\tfrac{\pi}{2} - \alpha - k\beta) \tag{11}$$

$$c = 2[\pi - 2\alpha + \tfrac{k^2}{3}(3 - 4k^2)^{\frac{1}{2}} - \beta k(3 - k^2)] \tag{12}$$

where *a* and β are:

$$\alpha = arc\ sen \frac{1}{2(1-k^2)^{\frac{1}{2}}} \qquad (13)$$

$$\beta = arc\ cos \frac{k}{[3(1-k^2)]^{\frac{1}{2}}} \qquad (14)$$

Finally, the new phase formed at the junction of the four grains has the following volume, Figure 6.

Figura 6 - Volume of the new phase formed by the meeting of four groups

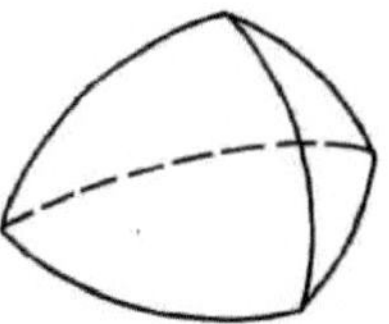

Source: CLEMM and FISHER (1955)

The coefficients shown are:

$$a = 3\{2\emptyset(1-k^2) - K[\left(1-k^2-\frac{K^2}{4}\right)^{\frac{1}{2}} - \frac{K}{\sqrt{8}}K]\} \qquad (15)$$

$$b = 24(\frac{\pi}{3} - k\emptyset - \delta) \qquad (16)$$

$$c = 2\{4\left(\frac{\pi}{3} - \delta\right) + kK\left|\left(1-k^2-\frac{K^2}{4}\right)^{\frac{1}{2}} - \frac{K}{\sqrt{8}}K\right| - 2k\emptyset(3-k^2)\} \qquad (17)$$

where *K*, 0 and δ are:

$$K = \frac{4}{3}\left(\frac{3}{2} - 2k^2\right)^{\frac{1}{2}} - \frac{2}{3}k \qquad (18)$$

$$\emptyset = arc\ sen \frac{K}{2(1-k^2)^{\frac{1}{2}}} \qquad (19)$$

$$\delta = arc\ cos \frac{\sqrt{2}-k(3-K^2)^{\frac{1}{2}}}{K(1-k^2)^{\frac{1}{2}}} \qquad (20)$$

To illustrate these equations, Clemm and Fisher (1955) generated a graph, shown in Figure 7, of the energy required to form a ferrite grain in an austenite matrix as a function of the radii of curvature of the ferrite nucleation surface nuclei for nucleation at the junction of two, three and four grains. The results clearly show that ferrite nucleation is more favorable at the four-grain junction.

Figura 7 - Ferrite nucleation energy as a function of surface curvature

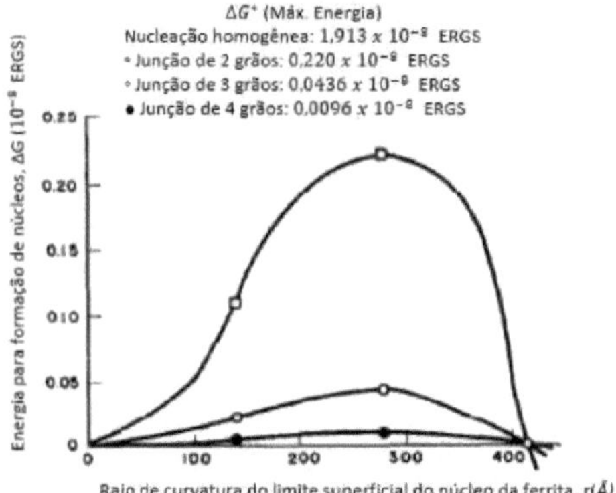

Source: Adapted from CLEMM and FISHER (1955)

2.1.3 Growth kinetics

Growth kinetics, or transformation kinetics, is generally described using the equation known as Johnson-Mehl-Avrami-Kolmogorov, JMAK. Growth kinetics has the characteristic of an "S" curve, a slow process initially, followed by an acceleration in growth and slowing down towards the end. This is due to an initial incubation period, which is the time to develop the first stable nuclei, followed by several stages of growth, which promotes an increase in the transformation rate. Finally, the transformation rate decreases as the microstructure approaches the end of transformation (Krauss, 2005). The equation that represents this behavior is:

$$f = 1 - e^{-kt^n} \tag{6}$$

where t is the time in seconds, k is a constant rate and n is known as the Avrami coefficient (Pereloma and Edmonds, 2012).

The central idea of the JMAK equation is to focus on the increment of the transformed volume fraction and relate it to the current value of the transformed fraction. Thus:

$$f = \frac{V}{V_{total}} \tag{7}$$

where f is the transformed fraction and V is the volume. To derive the equation, it is assumed that the homogeneous and heterogeneous nuclei are randomly distributed in space, isothermal crystallization conditions and growth rate independent of time. In this way, we can quantitatively relate the true increase in the transformed fraction (df), the extended fraction (df^{ext}), defined as the increase in the transformed fraction in relation to the initial volume of the untransformed material) and the current fraction:

$$df = df^{ext}(1-f) \quad (8)$$

Assuming also that the material has three-dimensional sphere-shaped nuclei of isotropic growth, we have:

$$V = \frac{4\pi}{3}r^3(t) = \frac{4\pi}{3}(vt)^3 \quad (9)$$

where r is the radius, t *is* the time and v is the growth rate. Having multiplied the individual volume by the density of nuclei, N, and obtained the increment of the extended fraction, we can substitute it into the differential equation (8).

$$f^{ext} = \frac{\Sigma V_i}{V_{total}} = \frac{4\pi}{3}N(vt)^3 \quad (10)$$

$$df^{ext} = \frac{V}{V_{total}} = 4\pi N v^3 t^2 dt \quad (11)$$

$$df = 4\pi N v^3 t^2 dt(1-f) \quad (12)$$

$$\frac{df}{(1-f)} = (4\pi N v^3)t^2 dt \quad (13)$$

By putting nucleation and growth in terms of a constant, which strongly depends on temperature:

$$k = \frac{4\pi}{3}Nv^3 \quad (14)$$

Solving the differential equation (13) and rearranging the equation gives us:

$$-\ln(1-f) = kt^3 \quad (15)$$

$$f = 1 - e^{-kt^n} \quad (16)$$

$$f = 1 - e^{-\frac{4\pi}{3}Nv^3t^3} \quad (17)$$

Equation (16) is a generic result, where n is related to the geometry of the transformation.

2.2 Transformation curves

From the JMAK equation described above, it is possible to obtain data on the fractions transformed from the solid material as a function of time at a constant temperature during the transformation. Taking the following example of the isothermal transformation of austenite into pearlite at 675°C, we obtain the first graph in Figure 8 in an "S" format where we have the percentage of transformation on the ordinate axis and the time in logarithmic form on the abscissa axis (CALLISTER, 2007, p. 348-349).

Figure 8 - Demonstration of how the transformation diagram is generated from the transformation percentage by time curve

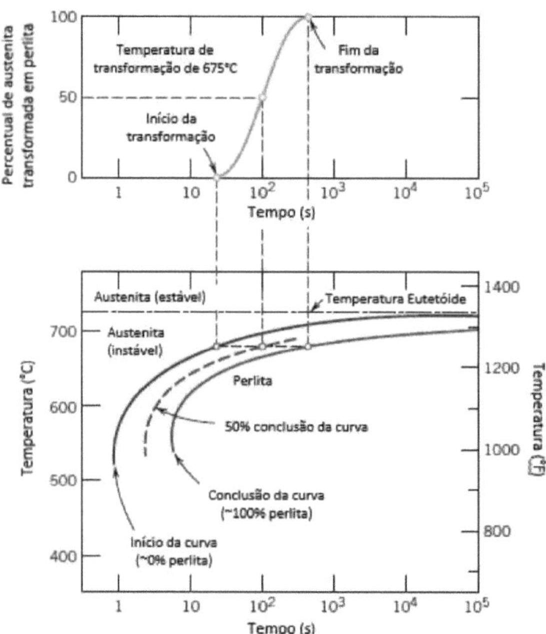

Source: Adapted from CALLISTER (2007)

The isothermal transformation curves have a similar shape to the one shown in Figure 9. This shape of the TTT curve is due to the fact that the temperature depends on the nucleation and growth rate of the grains. The construction of this isothermal curve can become quite complex as the possibilities for microstructure evolution increase. The microstructures typically found in steels are austenite, ferrite, cementite, bainite and martensite.

Figure 9 - Isothermal transformation

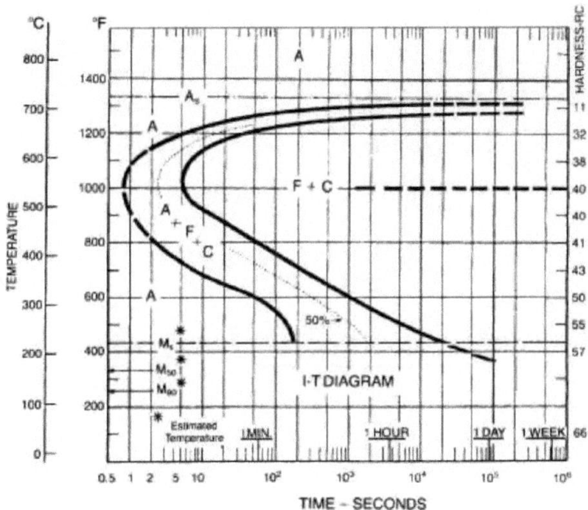

Source: Krauss (2005).

There are two ways of determining TTT curves experimentally, one using dilatometric techniques and the other metallography. Bearing in mind that the density of austenite is different from that of the new phases generated by its decomposition, the first technique consists of monitoring the change in volume over time of a sample cooled to the transformation temperature you want to study. The second technique uses the metallographic examination of several specimens that have been austenitized in the same way, time and temperature, and cooled simultaneously to the temperatures desired to evaluate the phase transformations, so that the evolution of the transformations can be monitored (COLPAERT, 2008, p. 202-203).

Continuous cooling transformation curves (TRC) are obtained using dilatometric techniques and are usually presented together with isothermal transformation curves (COLPAERT, 2008, p. 206). Figure 10 illustrates the TRC diagram superimposed on the TTT of a carbon steel alloy where there are two different cooling rates. This diagram has great instructive importance in predicting the continuous cooling treatments that can be obtained. Transformation begins when the cooling curve intersects with the reaction start curve and ends when it cuts the transformation curve. As shown in Figure 10, the microstructure obtained for the two cooling rates, fast and slow, was fine pearlite and coarse pearlite respectively. If the cooling curve does not cut the second transformation curve and passes between points A and B, the transformation from austenite to pearlite will cease when it intersects the martensite initiation curve, where the rest of the austenite that has not been transformed will be transformed into the martensitic microstructure (CALLISTER, 2007, p. 336337).

Figure 10 - Continuous cooling transformation diagram

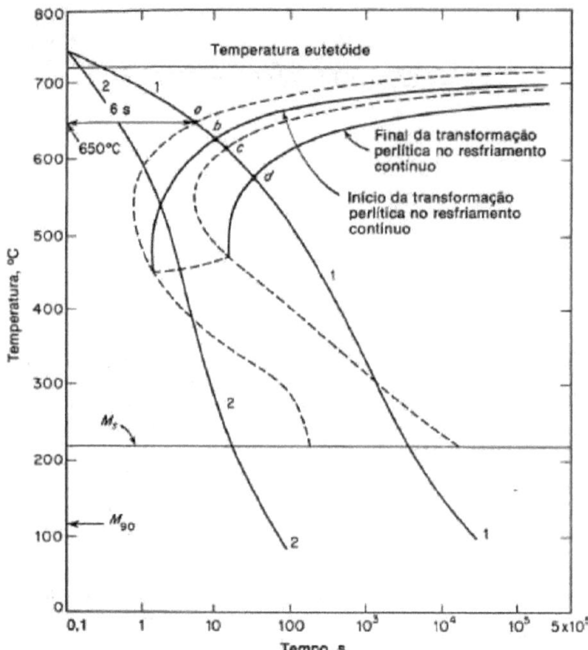

Source: Reed-Hill (1982)

2.3 Temperability

Hardenability is the variation in hardness of a steel from the surface to the core of the part when hardened. According to Reed-Hill and Abbaschian (2008), the composition and size of the austenitic grain, the severity of the tempering and the diameter of the bar are variables that are associated with the depth at which 50% martensite is obtained. In the case of steel composition, alloying elements will shift the perlitic transformation lines of a transformation diagram during continuous cooling. By shifting to the right, it can be said that the hardenability of the steel increases, as the martensitic structure can be obtained with a slower cooling rate, as shown in Figure 11. Observing the effect of grain size, the total number of nuclei formed varies directly with the available surface area, i.e. a fine-grained composite steel will have a larger grain boundary area than a coarse-grained steel and will therefore form pearlite more quickly, consequently having a lower hardenability (CHIAVERINI, 2008).

Figure 11 - Representation of the displacement of transformer lines

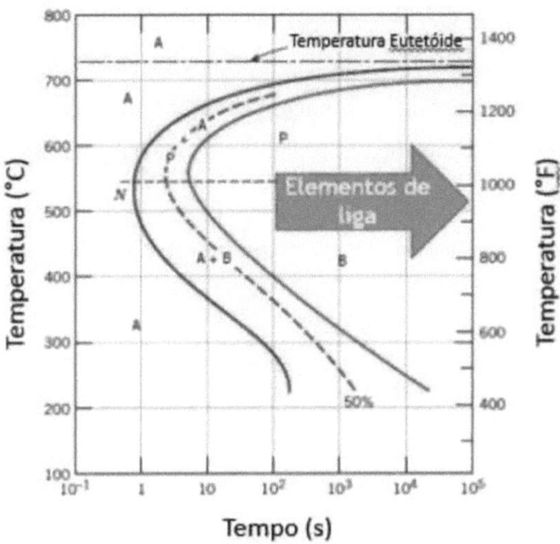

Source: Adapted from CALLISTER (2007)

Two methods have been established to quantify the hardenability of steels: Grossmann hardenability and Jominy hardenability. The Grossmann method uses the critical diameter and the hardening severity. Taking a cylindrical sample, a cut is made in the bar in order to analyze the circular cross-section to determine the hardness along the diameter. From the results obtained, a graph is generated (Figure 12), showing the variation in hardness along the diameter, with a higher hardness on the surface than in the center of the piece. At the inflection point of the graph, the microstructure is 50% martensite and 50% pearlite.

Figura 12 - Hardness curve along the hardened cylinder

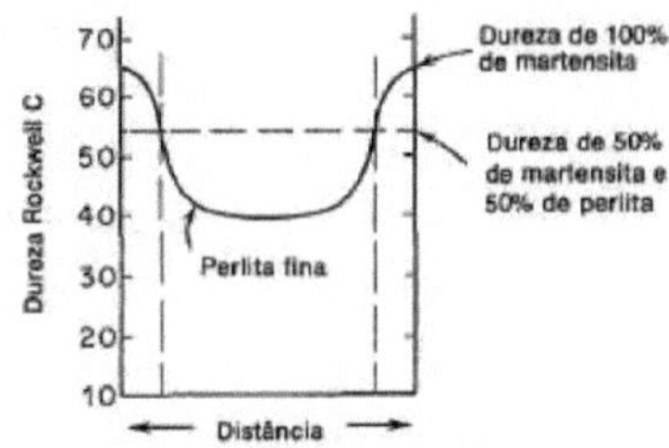

Source: Reed-Hill (1982)

This inflection point varies with the hardenability of the material. Considering several bars of the same steel with different diameters and undergoing the same hardening process, we can see how the variation in diameter influences the hardenability of the piece (Figure 13). Looking at the 25 mm bar, its center is 50% martensite and 50% pearlite, so all the bars with smaller diameters are fully hardened and those with larger diameters have a significant amount of pearlite in the center. This diameter is called the critical diameter and its value depends on the steel and the temper. The critical diameter of a steel is the measure of its tempering capacity, disregarding the cooling speed, since this variable is often referred to as a standard medium called ideal tempering, in which this hypothetical medium instantly assumes its temperature and remains at that value.

Figura 13 - Hardness curve for a series of steel bars of the same composition but different diameters

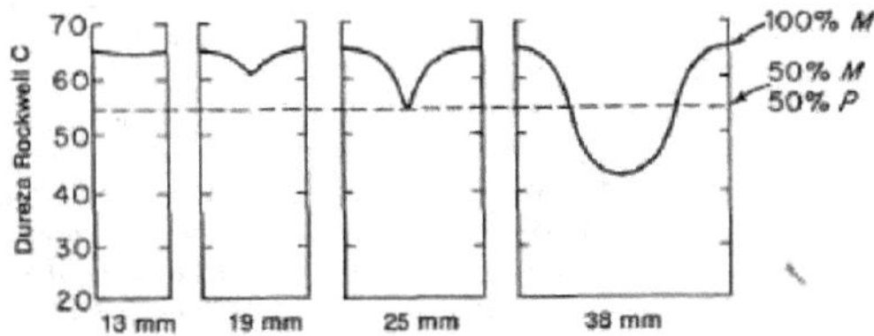

Source: Reed-Hill (1982)

Since the effect of the hardening medium is decisive, Table 1 was developed, which shows the hardening severity, or cooling speed, measured by the factor H. The higher the cooling speed, the higher the hardening severity, the greater the chance of distortion and cracks occurring in the material (COLPAERT, 2008, p. 286-287).

Table 1 - Severity of typical hardenings

H	*Condirai) de tèmpera*
0,20	Season in alcohol without stirring
0,35	Moderate oil quenching
0.50	Hardening at 0⅛Q with good agitation
0,70	Oil quenching with violent action
1,00	Hardening in water without stirring
1,50	Season with ⅛g□a with strong stirring
2,00	Season with SaJmoura without stirring
5,00	Brine quenching имя violent agitation
∞	Ideal tempering level

Source: Reed-Hill (1982)

To understand the use of hardening severity, the following example should be noted. If a bar has a critical diameter of 25 mm and has been cooled in brine without stirring (H=2.00), the ideal critical diameter will be 36 mm, i.e. the hardenability of the steel is DI = 36 mm (REED-HILL, 1982, p. 598-600). Thus, Figure 14 shows that if the piece is placed in a lower hardening severity, the ideal diameter will be less than 25 mm. The same applies to the opposite case, where H is greater than 2, the ideal diameter will be greater than 25 mm.

Figure 14 - Relationship between critical diameter and ideal critical diameter DI

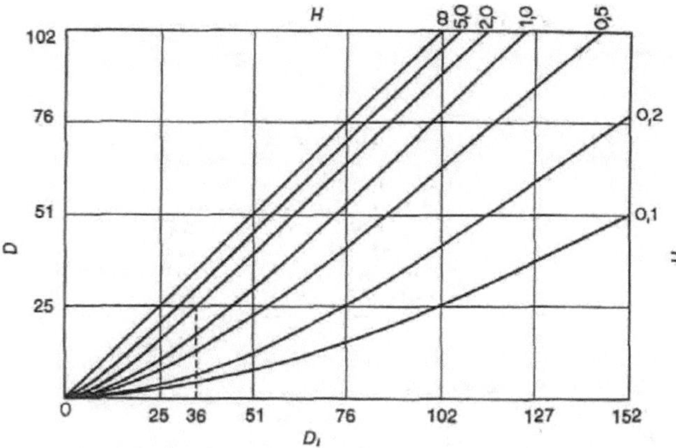

Source: Reed-Hill (1982)

According to Colpaert (2008), three stages occur when hot metal is placed in the liquid cooling medium of water. In the first stage, vapor bubbles form around the material, reducing the cooling speed as a result of the reduced heat flow between the part and the liquid. After the temperature drops a few hundred degrees, stage two begins and bubbles form and separate on the surface of the workpiece. At this point the cooling rate is maximum. In the last stage, conduction and convection take place.

Unlike the Grossmann method, the Jominy method is more convenient for determining hardenability due to the difficulty in carrying it out. In the Jominy test, a standard specimen is used in the form of a cylindrical bar 102 mm long and 25 mm in diameter. The piece is heated to the austenitizing temperature and held for a while in order to obtain a completely austenitic structure. The specimen is then taken to a support where it receives a constant jet of water on its lower base, which cools it quickly, while the upper part cools slowly in the air, as shown in Figure 15.

Figure 15 - Jominy hardenability test

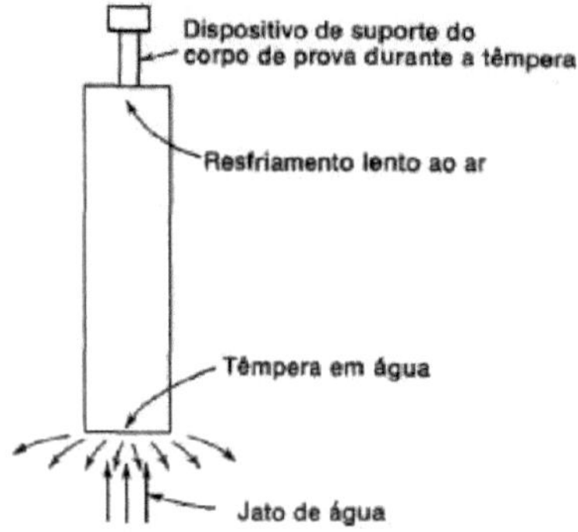

Source: Reed-Hill (1982)

Due to the different cooling rates along the length of the bar, it has different hardness values along its length and so the hardness variation curve is lifted. At the point where 50% of the martensitic structure is present, the distance to the end determines the ideal critical diameter. Figure 16 is an example of how the ideal critical diameter is determined. The hardness of the specimen 4 mm from the end shows 50% martensite, thus representing this ideal critical diameter.

Figure 16 - Hardness variation along the Jominy specimen

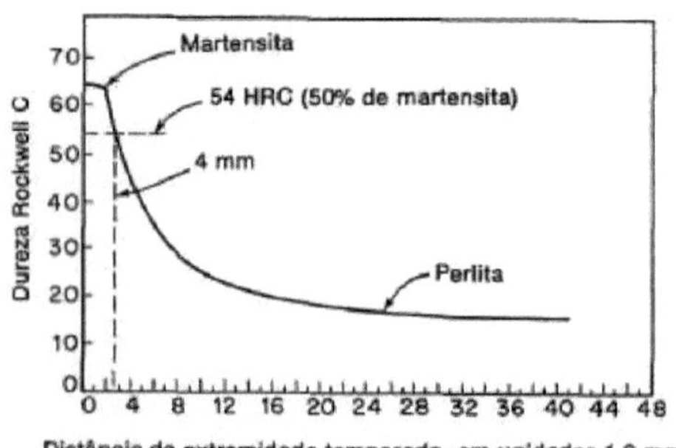

Source: Reed-Hill (1982)

3 LITERATURE REVIEW

3.1 Modeling isothermal transformation curves

Kirkaldy (1978) began his studies in the area of predicting TTT curves by trying to determine, through thermodynamic relationships, the temperature A and$_3$ as a function of the chemical composition of the steel alloy and the time required for the decomposition of austenite to be completed. The equation he proposed is linear for low amounts of alloying elements and strongly dependent on carbon. To prove this, the results obtained from the equation were compared with various experimental data obtained and it was concluded that for steels with less than 1% silicon and less than 7% total alloy, the prediction was excellent with errors of ±5% as expected. Generally speaking, the most suitable formula for predicting the Ae3 of a steel should have up to 1% Si, 2% Mo and 6% Cr, Ni, Mn and Cu and not exceed 10%. Values above this will present increasing errors due to approximations that are assumed in the mathematical formulation.

In 1980, Bhadeshia and Edmonds carried out an experiment in which they prepared an alloy steel containing 0.43% C, 3% Mn and 2.12% Si which was austenitized at 1200°C for 10 minutes and subsequently cooled to an isothermal transformation in a tin bath at 286°C for 30 minutes. After heat treatment, the microstructure observed was upper bainite. An analysis of the bainite formation mechanism using thermodynamic equations was carried out and the phenomenon of incomplete formation of the bainitic microstructure was understood.

Believing that in the near future it would be possible to construct the TTT diagrams of steels from any chemical composition using only theoretical calculations, Umemoto, Nishioka and Tamura (1981) studied the prediction from the TRC curves which provide information on the characteristics of the structures during cooling. The authors carried out their work on eutectoid steels where pearlite is the main diffusional transformation phase and presents only a "C" shaped curve in the TTT diagram. Using the kinetics of transformation for eutectoid steels, an equation was developed which determines the amount of pearlite transformed during continuous cooling.

Bearing in mind that the critical cooling rate is one of the fundamental parameters in determining the hardness of steels, it was defined that the slowest cooling rate that will produce a martensitic microstructure is called the upper critical cooling rate and the fastest cooling rate that will produce a structure without martensite is called the lower critical cooling rate. Thus Umemoto, Nishioka and Tamura (1981) developed equations from the transformation kinetics equation to determine the lower and upper critical cooling rates from the TTT diagrams. Similarly, using the transformation kinetics equation, they developed an equation to determine the distance in the Jominy bar from a cooling rate. Finally, based on the transformation kinetics equation, the ideal critical diameter was calculated from data obtained from TTT diagrams. The equations developed proved to be quite accurate compared to experimental data.

3.2 Theoretical evaluations of hardenability and applications

Victor Li et al. (1998) sought to refine previous work by developing a computer model for predicting microstructures during the heat treatment of steels, with hardness as the final result. The model was applied to predict the distribution of hardness along the Jominy bar, and five steps were determined for predicting Jominy hardness:

1) Using a thermodynamic model for heat-treatable steels;

2) Transient heat transfer analysis using two-dimensional finite elements for the Jominy bar;

3) Pre-calculating grain size based on an empirical model of austenite grain growth kinetics;

4) Calculating the microstructure evolution of the kinetic reaction model for austenite decomposition;

5) Calculating the distributed hardness along the Jominy bar based on empirical equations as a function of composition and cooling rate.

To validate their model, Victor Li et al. compared the hardness measured in the laboratory and its prediction using the computer model and observed that for low-alloy steels the model was very accurate. The major difference between the previous Kirkaldy model and this one was the reaction rate equation which affects the transformation phase. Figure 17 and Figure 18 compare the proposed model, the Kirkaldy model and real measurements, so it was possible to see that the model corrected problems related to previous models.

Figure 17 - Predicted and measured Jominy hardness for ASTM A588 steel

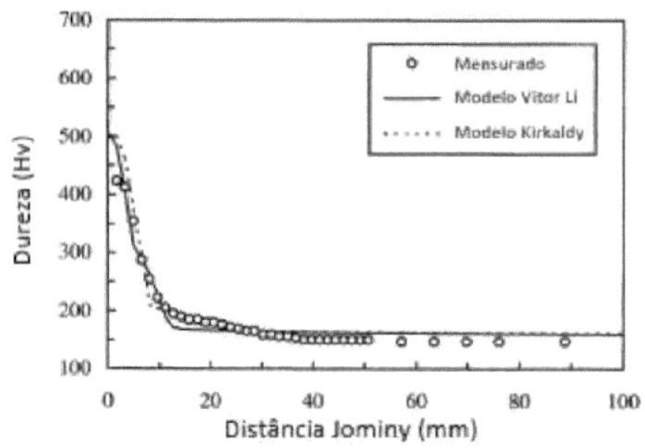

Source: Adapted by Vitor Li et al (1998)

Figure 18 - Predicted and measured Jominy hardness for AISI 4140 steel

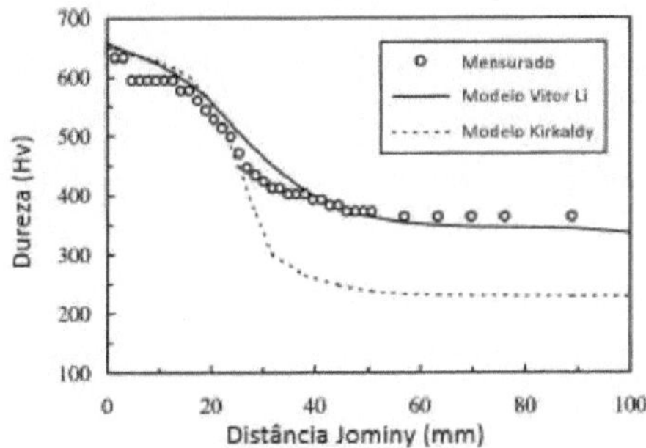

Source: Adapted by Vitor Li et al (1998)

In 2009, Guo et al. developed a computational model for predicting distortions during heat treatment. For this study, it was necessary to look at previous studies of the material properties of each phase formed during heat treatment, which is dependent on the chemical composition of the alloy. Thus, the use of TTT and TRC diagrams, thermophysical properties and mechanical properties was necessary to predict the distortions.

To demonstrate how material properties change with the cooling rate, Guo used a 4140 steel alloy with two cooling rates, 20°C/s and 5°C/s, and observed that for the first cooling rate there was a large formation of martensite while at the 5°C/s rate there was a much lower appearance of martensite than the first. Consequently, properties relevant to the prediction of distortion, such as density, coefficient of linear expansion, thermal conductivity and subjected stress, are affected. The success of the model is based on the accuracy of the property calculations during the phase transformations.

According to another methodology for predicting the continuous cooling diagram, Trzaska and Dobrzanski in 2007 used a hybrid model to calculate the austenite transformation during continuous cooling. Regression analysis, neural networks and a collection of empirical data were used. The neural networks applied to the calculations provided a great deal of correction, but showed a difference between the calculated and empirical values.

In 2013, Trzaska again developed a new hybrid model combining several mathematical tools including logistic regression and multiple regression. For the calculation of hardness, neural networks were used and an empirical database of profound significance was used to prepare the hardness calculation method. The model developed allowed the hardness of the steel to be calculated from the chemical composition. In both of Trzaska's articles, the austenite grain size and the austenitizing time were not taken into account due to the lack of information in most of the continuous cooling diagrams used.

The prediction of phase transformations of microstructures and mechanical properties has been studied in various ways, one of which is the prediction of austenite decomposition kinetics, in which researchers have combined the heat transfer problem with the transformation problem. In 2004, Serajzadeh and Taheri worked on a model based on the finite element method and assuming a transformation with a second degree equation for predicting the kinetics of austenite decomposition during continuous cooling for low carbon steels. To deal with the temperature dependence of the material property during cooling, an iterative procedure was employed due to the non-linear nature of the problem. To validate the results of the model, the prediction was compared with the historical time-temperature curves and the microstructures.

Carlone, Palazzo and Pasquino, (2010), studied the temperature transient during solid-solid phase transformations throughout the heating, homogenization and cooling process by modeling the formation and decomposition of austenite, taking into account nucleation and grain growth. The results obtained for the continuous heating simulation were validated by comparing the results obtained in previous articles. For continuous cooling, the simulation was validated against experimental data from the continuous cooling diagram, which showed a good approximation to the real values. Finally, a finite element simulation of the cooling of a cylinder in water was carried out in order to evaluate various cooling rates along the cylinder, thus obtaining various hardness values along the part due to the different structures generated. The latter also proved to be in line with previous studies.

Sushanthi and Maity, (2014. 2015), also developed a model by solving heat transfer equations using finite difference. For the hardness prediction, the TTT diagram was used instead of the TRC diagram, due to the fact that the percentage transformation difference can be determined experimentally with high precision. The predicted critical diameter was very close to the experimental value, with only a 6% deviation.

Chen et al. in 2014 studied the decomposition of austenite during continuous cooling of a 22MnB5 boron steel. Their study used finite element analysis to predict the influence of temperature during phase transformation. The JMAK (Johnson-Mehl-Avrami-Kolmogorov) equations were used to describe the decomposition of austenite into ferrite, pearlite and bainite during continuous cooling. For martensite transformation, the K-M (Koistinen and Marburger) model adjusted by experiments was used. The simulated results obtained were in agreement with the experimental values, demonstrating that the JMAK equations can provide a good prediction of transformation kinetics. Like the K-M model, the simulated results for TRC show a good ability to simulate austenite decomposition.

(2015) also focused on transformation kinetics. Their subject of study was the prediction of phase transformation and hardness of a boron-alloyed low-carbon steel 22MnB5. The initial model successfully describes the isothermal transformation and provides the isothermal transformation diagram. However, the temperature of the workpiece under study is constantly changing, making it necessary to convert the isothermal model into a non-isothermal model of the transformation kinetics during cooling. In order to validate the performance of the phase transformation model, various cooling rates corresponding to the expansion experiments were simulated and the results were shown to be in agreement with the experiments.

In 2016, Chen *et al.* developed a new study presenting an inverse model by constructing a mathematical relationship between the kinetics of continuous cooling and the isothermal transformation curve. The JMAK parameters were deduced from the experimental TRC kinetics. The experimental steel used was G17CrMo9-10 with a completely bainitic microstructure. The prediction for the transformation time of the TTT diagram using the TRC showed a good result for isothermal temperature regions of a completely bainitic transformation. Thus, the reliability of the proposed model was validated using dilatometry on the completely bainitic steel alloy.

In 2015, Nunura developed his work in a different way to the researchers mentioned above. He modeled the Jominy curve by experimentally acquiring the curves and cooling rates along a Jominy bar and obtained numerical expressions that correlate the percentage of each phase and the hardness as a function of the cooling rates. In his study, Nunura observed that the morphology of the phases obtained changed according to the austenitizing temperature and the higher the temperature, the higher the Jominy hardness profile. The hardness values calculated from the material composition are in agreement with the experimental results.

With the development of computer models capable of predicting the microstructure formed in steels as a function of cooling conditions, researchers have recently turned their attention to the development of manufacturing process simulations capable of taking into account phase changes that occur during the manufacturing stages. For example, in heat treatment processes (DOMANSKI and BOKOTA, 2011), welding (MAZAR ATABAKI *et al.*, 2016) and hot mechanical forming processes, notably the stamping of automotive panels (HAGENAH *et al.*, 2015).

In welding, the heterogeneity of heat distribution along the joint means that there is a volumetric incompatibility between the effects of thermal expansion and contraction. These incompatibilities are accommodated by elastic deformations in the part, generating high residual stresses or even distortions or warping. Phase changes during cooling can also contribute to the evolution of residual stresses, since the formation of martensite from austenite (a more compact crystalline arrangement) is associated with volumetric expansion in the material. In this way, ways have been sought to incorporate models capable of describing the expected phase changes into virtual models of the cooling of welded components, obtaining a more realistic description of the state of stress produced by the welding process (O'MEARA, *et al.*, 2015; HAMELIN *et al.*, 2014; WANG *et al.*, 2017).

In hot stamping processes, a previously austenitized steel blank is formed using a matrix at room temperature, which can lead to the formation of different constituents such as pearlite, bainite and martensite. Predicting the relative fractions of the constituents obtained and, consequently, the mechanical properties of the final product, depends on a joint analysis of the cooling during the forming process and the resulting phase transformations. Faced with the prospect of obtaining stamped parts with controlled properties, researchers have sought to couple the problem of cooling analysis during stamping with microstructure prediction models (KO *et al.*, 2015; HIPPCHEN *et al.*, 2016).

4 METHODOLOGY

4.1 Numerical Methodology

Vitor Li's (1998) model for the TTT diagram was taken as the initial basis for this work. This work was chosen as a basis due to its wide range of citations in other works, the level of detail and the ease of its computer implementation. The model implemented on the basis of the original equations (VITOR LI, 1998) then served as a platform for testing strategies for correcting the modeled TTT diagram by comparing it with experimental Jominy hardenability data for selected steels. All programming was carried out in the C++ language. Figure 19 shows the flowchart illustrating the principle of operation of the *software* developed. The comments and equations describing each step of the flowchart will be presented below.

Figure 19 - Flowchart of the algorithm

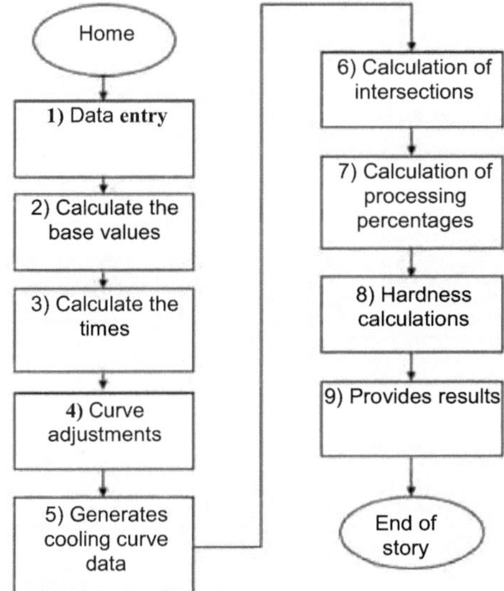

Source: Prepared by the author

Use of the *software* developed in this work begins with information on some of the constants used in the equations, such as the grain size (standard value of G=7), the gas constant (standard value of R = 1.98 [illegible]) and the activation energy of the diffusion reaction (standard value of Q = 27500 [illegible]). This is followed by data entry (number 1 in the flowchart) relating to the chemical composition (% weight) of the steel to be analyzed. Figure 20 shows the screen of the *software* developed for entering the aforementioned data. In addition to the fields for information on the steel's chemical composition and the physical constants to be considered, the reference cooling rates and the experimental hardness points to be used for comparing the theoretical and real Jominy hardenability curves.

Figure 20 - Data entry screen

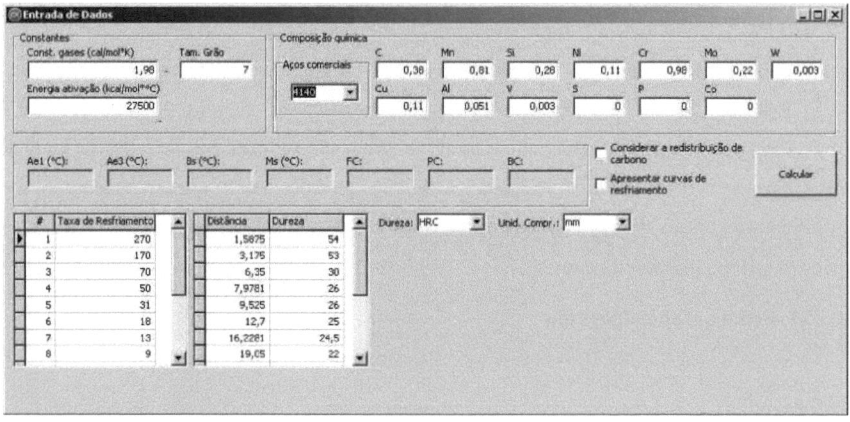

Source: Prepared by the author

From the general formula for austenite decomposition, equation 18, the equations for the times of the TTT curves and the transformation percentages of each microstructure are derived (VICTOR LI *et al.*, 1998).

$$\tau(X,T) = \frac{F(C,Mn,Si,Ni,Cr,Mo,G)}{\Delta T^{n}\exp\left(-\frac{Q}{RT}\right)}S(X) \tag{18}$$

For the transformation kinetics model of austenite decomposition, it is necessary to determine the reaction rate term (VICTOR LI *et al.*, 1998):

$$S(X) = \int_0^X \frac{dX}{X^{0,4(1-X)}(1-X)^{0,4X}} \tag{19}$$

From Equation 19 it is necessary to determine two reaction rate values, one for an initial value (with the integral ranging from 0 to 0.001) and a final value (with the integral ranging from 0 to 0.999). These values represent the start and end of the transformation of each of the following equations.

From the composition, it is possible to calculate the times for each temperature from the following equations, the first being the ferrite reaction represented by (VICTOR LI *et al.*, 1998):

$$\tau_F = \frac{FC}{2^{0,41G}(Ae_3-T)^3\exp\left(-\frac{Q}{R(T+273)}\right)}S(X) \tag{20}$$

where FC and A and$_3$ are directly affected by the chemical composition:

$$FC = \exp(1 + 6{,}31C + 1{,}78Mn + 0{,}31Si + 1{,}12Ni + 2{,}70Cr + 4{,}06Mo) \quad (21)$$

$$Ae_3 = 937{,}3 - 224{,}5C^{0,5} - 17Mn + 34Si - 14Ni + 21{,}6Mo + 41{,}8V - 20Cu \quad (22)$$

The pearlite equation is represented by (VICTOR LI *et al.*, 1998):

$$\tau_P = \frac{PC}{2^{0,32G}(Ae_1 - T)^3 \exp\left(-\frac{Q}{R(T+273)}\right)} S(X) \quad (23)$$

where PC and Ae_1 are given by:

$$PC = exp(-4{,}25 + 4{,}12C + 4{,}36Mn + 0{,}44Si + 1{,}71Ni + 3{,}33Cr + 5{,}19\sqrt{Mo}) \quad (24)$$

$$Ae_1 = 739{,}3 - 22{,}8C - 6{,}8Mn + 18{,}2Si + 11{,}7Cr - 15Ni - 6{,}4Mo - 5V - 28Cu \quad (25)$$

Finally, the bainite equation is given by (VICTOR LI *et al.*, 1998):

$$\tau_B = \frac{BC}{2^{0,29G}(B_S - T)^2 \exp\left(-\frac{Q}{R(T+273)}\right)} S(X) \quad (26)$$

where BC and Bs given by:

$$BC = exp(-10{,}23 + 10{,}18C + 0{,}85Mn + 0{,}55Ni + 0{,}90Cr + 0{,}36Mo) \quad (27)$$

$$B_s = 637 - 58C - 35Mn - 15Ni - 34Cr - 41Mo \quad (28)$$

equations A and_1 , A and_3 and B_s are given in °C and represent the lower equilibrium temperature between bainite, ferrite and austenite, the upper temperature between ferrite and austenite and the initial isothermal temperature for the formation of austenite-bainite on cooling.

Finally, it is necessary to have the martensite transformation start temperature, which is given in °C (KUNG and RAYMENT, 1982).

$$M_S = 539 - 423C - 30{,}4Mn - 17{,}7Ni - 12{,}1Cr - 7{,}5Mo + 10Co - 7{,}5Si \quad (29)$$

The values of S(x), FC, PC, BC, Ae3, Ae1, Bs and Ms are the values determined in step two of the flowchart and the times of each microstructure are represented by step three.

With these equations, the iterations are started, varying from 1 to 900°C, to calculate the start and end times for the decomposition of austenite into the constituents pro-eutectoid ferrite, pearlite and bainite. The temperature of 900°C was determined as the limit because at this temperature a hypoeutectoid steel is completely situated in the austenitic phase field. The adjustment of the curves explained in the flowchart (number 4) is due to the fact that the curves are limited to the meeting between them and do not "cross", for example the ferrite curve passes through the bainite curve.

To determine the percentage of transformation in each phase, it is necessary to have the cooling curves superimposed on the transformation curves. To generate the cooling curves (number 5 in the flowchart), the cooling starts at 900°C and for each cooling rate the temperature decreases until it reaches 0°C. Figure 21 shows the screen with the results panels of the *software* developed, keeping the representations of the cooling curves (which may or may not be displayed). In the panel shown in Figure 21, the user can, in addition to viewing the TTT diagram generated, export the data obtained and view the Jominy hardenability diagram of the steel selected for analysis.

Figure 21 - Visualization of cooling curves superimposed on the modeled transformation diagram for ABNT 4140 steel

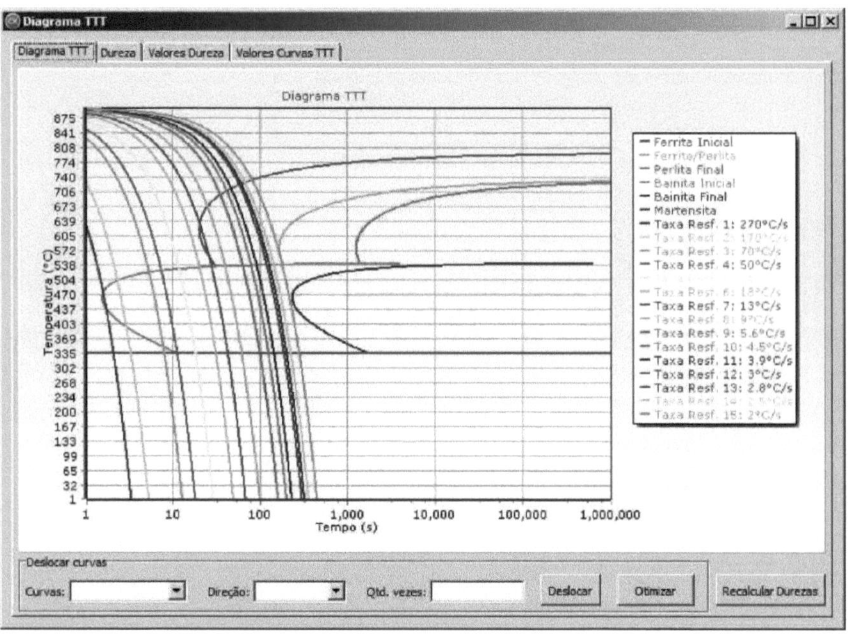

Source: Prepared by the author

Step six of the flowchart is the calculation of the intersections between the two sets of curves indicating the points (time and temperature) relative to the start and end of each phase transformation of the steel. A proprietary algorithm was used to calculate the intersections. Once the intersection points have been determined, the equations for calculating each percentage of each transformed phase are derived from equation (30) (number 7 in the flowchart).

$$X = \int_0^t \frac{\Delta T^n e^{-\frac{Q}{RT}} X^{0,4(1-X)} (1-X)^{0,4X}}{F(C, Mn, Si, Ni, Cr, Mo, G)} \tag{30}$$

The equations obtained for each type of microstructure, ferrite, pearlite and bainite, are respectively as follows:

- Ferrite:

$$X = \int_0^t \frac{2^{0,41G}(Ae_3 - T)^3 e^{-\frac{Q}{R(T+273)}} X^{0,4(1-X)}(1-X)^{0,4X}}{FC} \quad (31)$$

- Pearlite:

$$X = \int_0^t \frac{2^{0,32G}(Ae_1 - T)^3 e^{-\frac{Q}{R(T+273)}} X^{0,4(1-X)}(1-X)^{0,4X}}{FC} \quad (32)$$

- Bainite:

$$X = \int_0^t \frac{2^{0,29G}(B_S - T)^2 e^{-\frac{Q}{R(T+273)}} X^{0,4(1-X)}(1-X)^{0,4X}}{FC} \quad (33)$$

In this work, equations (20), (23) and (26) allow the isothermal transformation curves of a steel to be calculated. However, continuous cooling rates are considered for the purpose of determining the fraction of phases present in the material. To solve this problem, the cooling rates to be used to determine the phase fractions using equations (31), (32) and (33) were discretized into steps involving alternately maintaining a constant temperature over time and a drop in temperature without varying the time. In other words, the calculation of the fraction of phases will be determined by adopting Scheil's addition rule, as illustrated in Figure 22. Each of the equations (31), (32) and (33) is used according to the constituents being formed in the integration step.

Figure 22 - Illustration of the procedure to be adopted for converting continuous cooling rates into cooling curves with isothermal steps

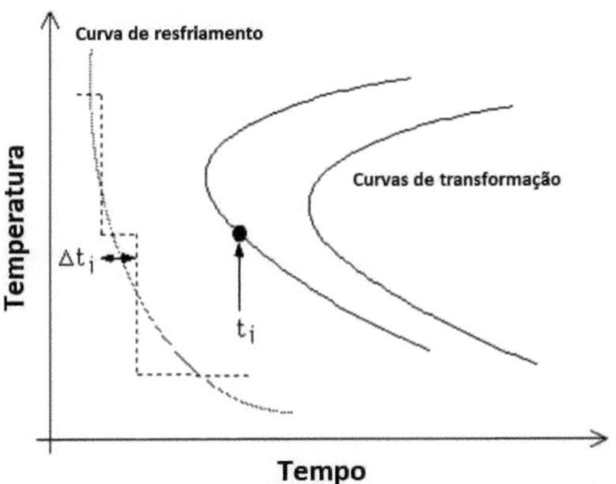

Source: Adapted from Bhadeshia (2017)

By using cooling curves equivalent to those observed along the Jominy bar in a hardenability test, it is possible to obtain the transformation percentages of each phase and, consequently, it is possible to calculate the distributed hardness (number eight on the flow chart) along a Jominy bar (or for any previously established cooling rates) using the mixing rule (STEVEN and HAYNES *apud* VICTOR LI *et al.*, 1998):

$$H_V = X_M H_{V_M} + X_B H_{V_B} + (X_F + X_P) H_{V_{F+P}} \tag{31}$$

where H_V is the hardness in Vickers and X_M , X_B, X_F and X_P are the transformed volume fractions of martensite, bainite, ferrite and pearlite respectively, which are obtained by integrating each curve limited by the intersections obtained. Hv_M , Hv_B and $Hv_{(F+P)}$) are the hardnesses of martensite, bainite and the mixture of ferrite and pearlite and they are functions of the steel composition and the cooling rate (STEVEN and HAYNES *apud* VICTOR LI *et al.*, 1998):

$$H_{V_M} = 127 + 949C + 27Si + 11Mn + 8Ni + 16Cr + 21 \log V_r \tag{32}$$

$$H_{V_B} = -323 + 185C + 330Si + 153Mn + 65Ni + 144Cr + 191Mo + (89 + 53C - 55Si - 22Mn - 10Ni - 20Cr - 33Mo) \log V_r \tag{33}$$

$$H_{V_{F+P}} = 42 + 223C + 53Si + 30Mn + 12{,}6Ni + 7Cr + 19Mo + (10 - 19Si + 4Ni + 8Cr + 130V) \log V_r \tag{34}$$

where V_r is the cooling rate at 700°C per hour. Each hardness value is obtained as a function of the cooling rate used. After all the calculations have been made, the last step of the flowchart (number nine) is to display the results on screen for the user, which are the TTT and Jominy diagrams.

Once the expected hardnesses along the Jominy bar have been determined from the virtual transformation curves, a comparison is made between the theoretical and experimental hardenability curves. To this end, an interactive optimization algorithm has been programmed which, based on the user's selection, allows an isothermal transformation curve to be shifted in time. With the modification generated in the TTT diagram, the Jominy curve is recalculated. The parameter used in the optimization is the average absolute deviation between the experimental Jominy curves.

4.1.1 Calculation of the redistribution of carbon and residual austenite

In section 4.1, the description of the implementation of the model for generating the TTT and Jominy hardenability diagrams followed the steps outlined in the work by Victor Li *et al.* (1998). As a way of seeking possible refinement in the calculations, the carbon redistribution and residual austenite calculations were also implemented.

The formation of ferrite from austenite during the cooling of a steel increases the carbon content of the austenite that has not yet been transformed, since ferrite is capable of dissolving a very small amount of carbon compared to austenite. As the austenite that exists in the material with the formation of ferrite is gradually enriched with carbon, then the calculation of the hardness of the martensite formed (from this modified austenite) must also be altered. Thus, the correction is justified not only because martensite is formed perfectly (a fraction of untransformed austenite remains, the residual austenite) but also because the martensite formed has more carbon in solution than the amount initially predicted by the steel's chemical composition.

The calculation of the correction of the carbon content present in the austenite as a function of the ferrite fraction formed during cooling was implemented according to expression 35, below:

$$C_\gamma = \frac{0{,}022F_\alpha - C_x}{F_\alpha - 1} \qquad (35)$$

where C_γ is the amount of carbon in the austenite, F_a is the fraction of ferrite formed in the steel and Cx is the amount of carbon in the steel (informed by the user). The constant 0.022 is obtained from the equilibrium diagram and represents the limit of carbon solubility in the ferrite. From this new carbon value for each percentage of ferrite formed, a new Ms value was calculated using formula 29 and the new carbon values calculated in formula 35. The residual austenite fraction was then calculated using the formula:

$$F_\gamma = e^{(-0{,}011(M_s - 25))} \qquad (36)$$

As there is an Ms value for each cooling time/temperature, there is a corresponding value for the residual austenite fraction. This formula is due to Koistinen and Marburger (1958) who analyzed the austenite-martensite transformation process and noted the redistribution of carbon during the formation of ferrite. The proposed calculation would be valid for cooling under equilibrium conditions, which is an important simplification for the case proposed in this study. Therefore, the implementation of the carbon redistribution calculation was presented in the *software* developed as an option for the user.

4.2 Experimental Methodology

In order to verify the model generated for the TTT curves and also the algorithm developed to adjust these curves based on the Jominy test results, hardenability tests will be carried out on samples of AISI 1045, 4140 and 4340 steels. The nominal chemical composition of the steels is shown in Table 2. The geometry of the specimens for the hardenability test is shown in Figure 23.

Table 2 - Chemical Composition of 1045, 4140 and 4340 Steels (% by weight)

Aço	C	Mn	Si	Ni	Cr	Mo	W	Cu	Al	V	S	P
1045	0,48	0,75	0,35	-	-	-	-	-	-	-	0,05	0,04
4140	0,38	0,81	0,28	0,11	0,98	0,22	0,003	0,11	0,05	0,003	-	-
4340	0,41	0,74	0,25	1,72	0,77	0,27	0,007	0,07	0,03	0,002	-	-

Figure 23 - Jominy specimen dimensions

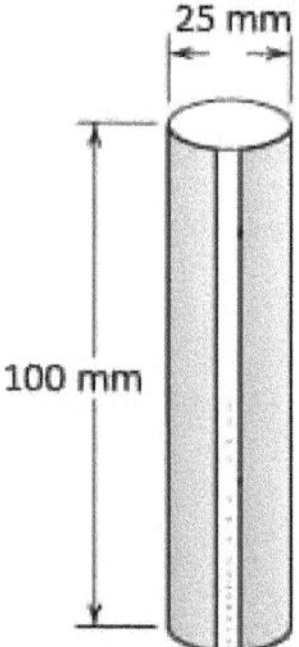

Source: Callister (2007)

The steels were chosen because of their widespread use in metal-mechanical industries for manufacturing various components. The hardenability tests, including the heating and cooling stages, will be carried out in the Mechanical Construction Materials laboratory at PUC Minas. Hardness measurements along the axis of the specimen will be carried out using the Rockwell C technique.

5 RESULTS AND DISCUSSION

The steels analyzed were ABNT 1045, 4140 and 4340 and the results obtained are analyzed in this chapter as follows: the diagrams generated by the model, the Jominy curves generated by the model compared with the experimental curves and, finally, the same results obtained after executing the optimization routine developed.

5.1 Model diagrams

On the data entry screen, each of the steels was selected to run the algorithm without displaying the cooling curves. Figure 24 below shows the TTT diagram for AISI 1045 steel and Figure 25 compares the real diagram with the diagram generated by the algorithm. The same data is shown for ABNT 4140 and 4340 steels in Figures 26 to 30, respectively.

Figure 24 - Modeled TTT diagram of 1045 steel

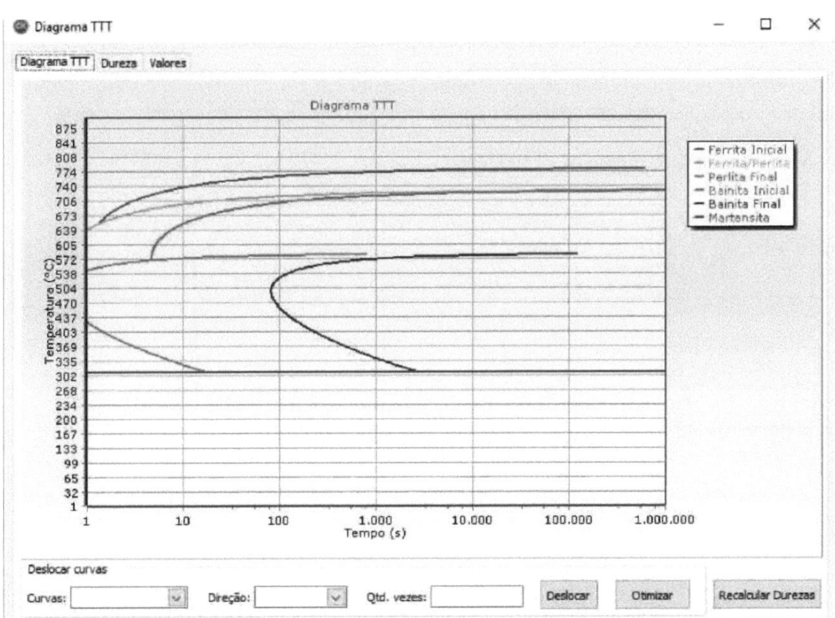

Source: Prepared by the author

Figure 25 - Modeled and real TTT diagram of 1045 steel

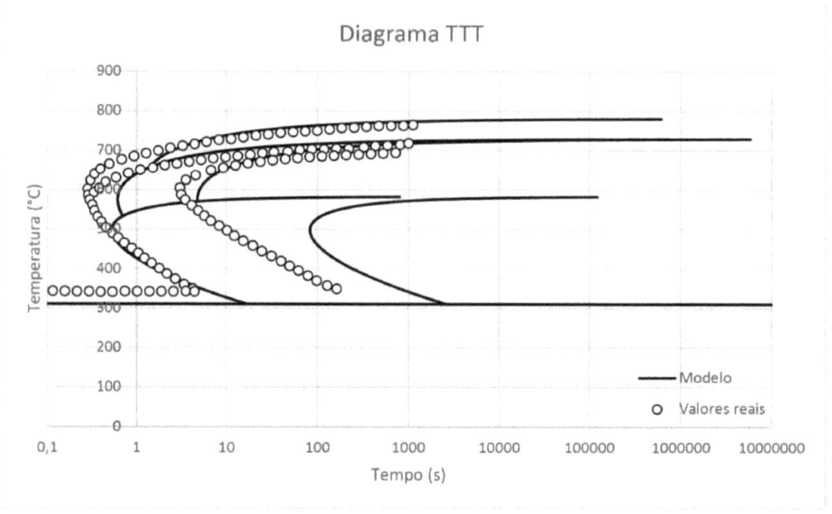

Source: Prepared by the author. Experimental data from the TTT diagram obtained from Callister (2007).

Figure 26 - Modeled TTT diagram of 4140 steel

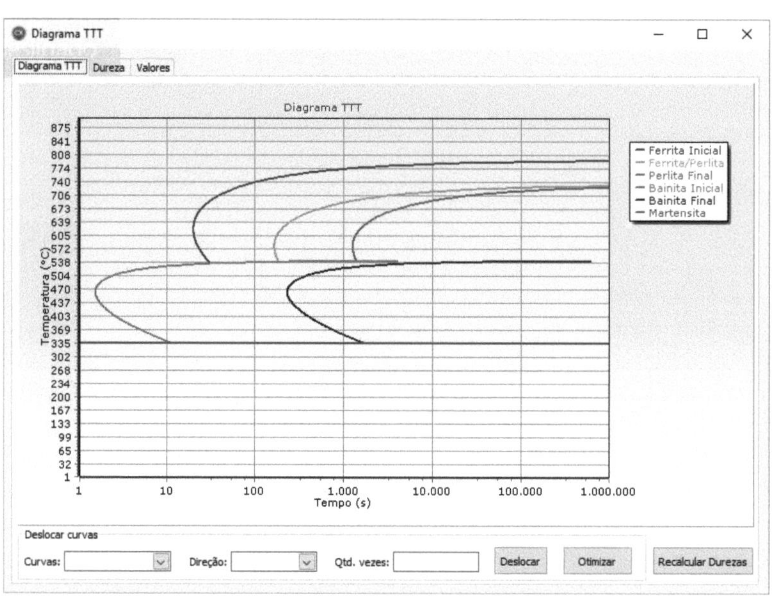

Source: Prepared by the author

Figure 27 - Modeled and real TTT diagram of 4340 steel

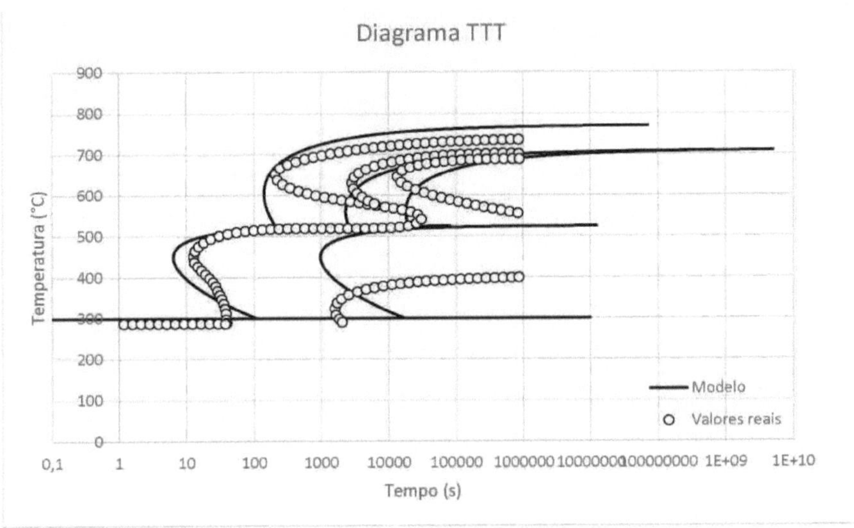

Source: Prepared by the author

Figure 28 - Modeled TTT diagram of 4340 steel

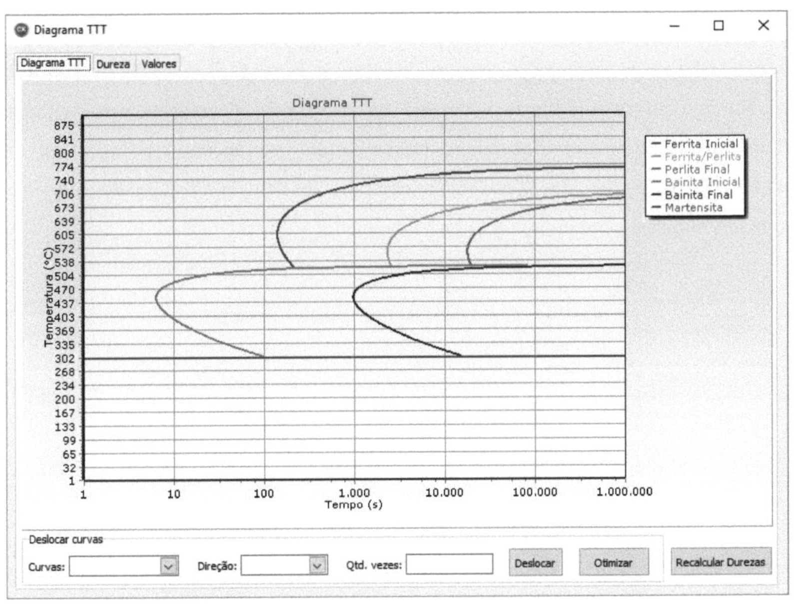

Source: Prepared by the author

Figure 29 - Modeled and real TTT diagram for 4340 steel

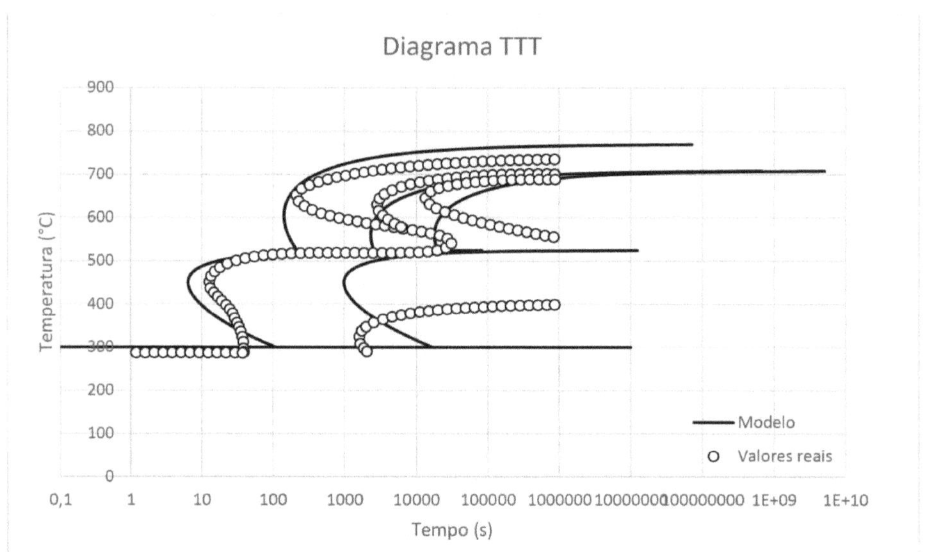

Source: Prepared by the author

5.2 Jominy curves

In the same sequence as the steels presented in the previous section, a comparison was made between the experimental Jominy curves and those generated by the model developed, first without considering carbon redistribution and then taking redistribution into account. Using a set of experimental data for AISI 1045 steel, it was possible to compare this data with that generated by the model. Figure 30 below shows a good approximation of the model with real data.

Figure 30 - Jominy diagram 1045 steel

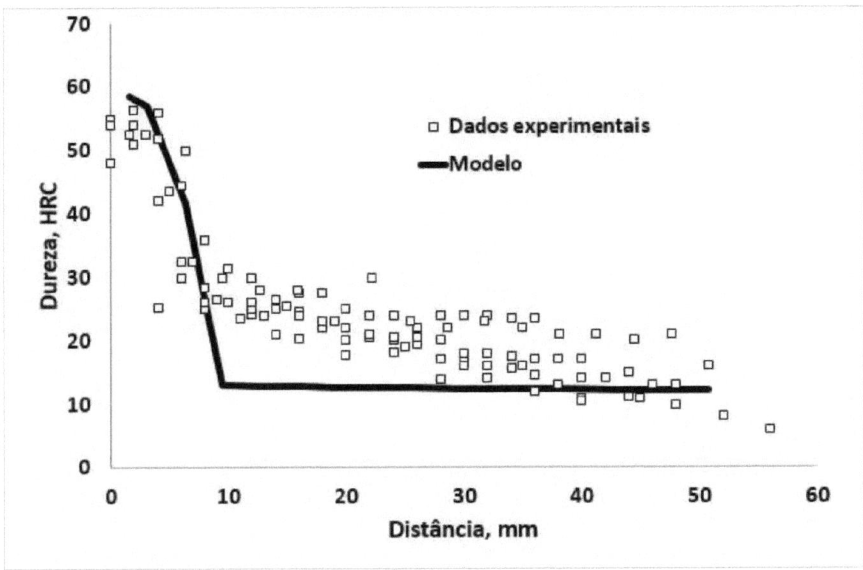

Source: Prepared by the author

Figure 31 shows the distribution of the microstructure generated by the model along the length. From each of these transformed percentages obtained, it is possible to calculate the hardness along the length of the structure. It can be seen that the sharp drop in the martensitic microstructure combined with an increase in the pearlitic microstructure resulted in a rapid drop in the hardness value of the material. Similar results were obtained by Nunura *et al* (2015), who studied the variation in the fraction of microconstituents along the Jominy bar, and noted the elimination of martensite for distances greater than 6.36 mm from the end of the hardened bar. In addition, it can be seen that the values for the fractions of ferrite and pearlite constituents for slow cooling rates gradually approach the values expected for cooling under equilibrium conditions (42% and 58% for ferrite and pearlite, respectively). The high values observed for the pearlite fraction at distances close to the tempered end are also consistent with experimental observations, as can be seen by analyzing Figure 32(a) and (b), which shows the microstructures observed at distances of 10 and 20 mm from the tempered end.

Figure 31 - Transformed percentages 1045 steel

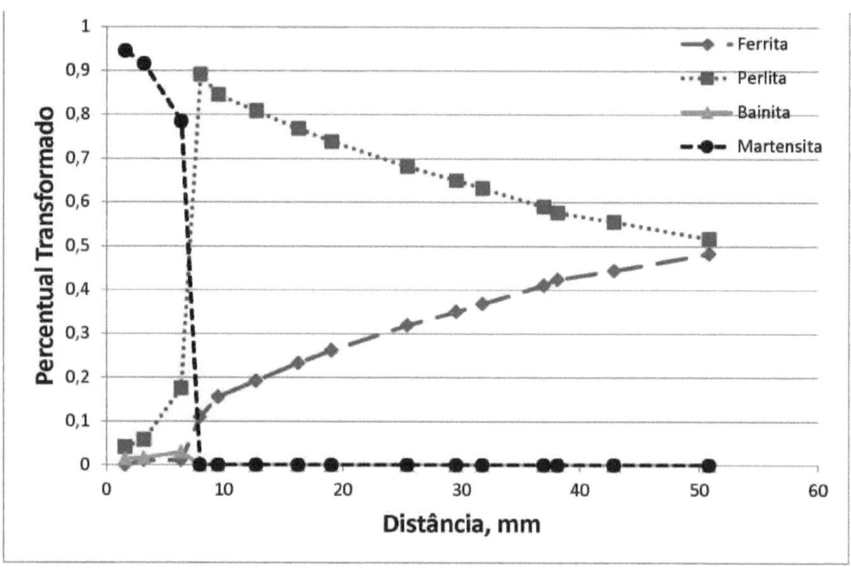

Source: Prepared by the author

Figure 32 - Typical microstructures obtained at distances of (a) 10 mm and (b) 20 mm from the hardened end

(a) (b)

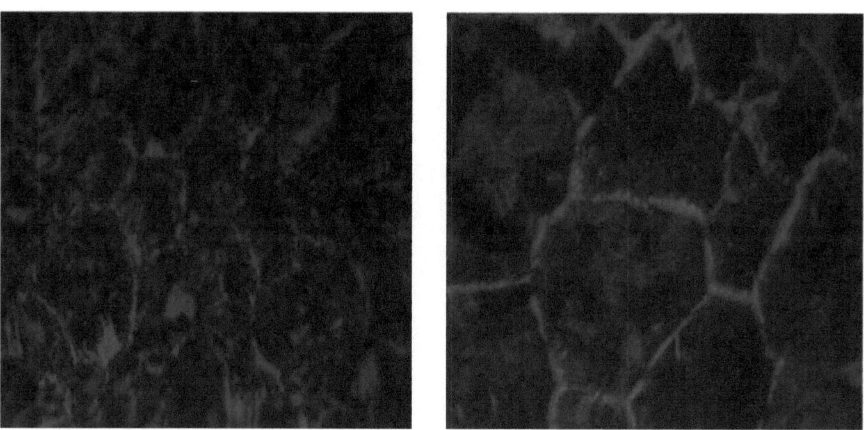

Source: Prepared by the author

Making the same comparison as before, but taking into account the redistribution of carbon in the austenite caused by the formation of ferrite and the possible effects on the accumulation of residual austenite, it was not possible to notice any significant changes in the hardness evolution of the material (Figure 33). However, analysis of the fractions of the constituents as a function of distance from the hardened end of the Jominy bar reveals the fragility of applying the model adopted for calculating residual austenite to this case (Figure 34), due to the unrealistic values obtained: close to 50% for distances equivalent to half the length of the Jominy bar. Despite this, it can be seen that the trend itself, of increasing residual austenite as the cooling rate decreases, is consistent with empirical knowledge (see, for example, Bangaru and Sachdev, 1982).

Figure 33 - Jominy diagram of 1045 steel considering carbon redistribution

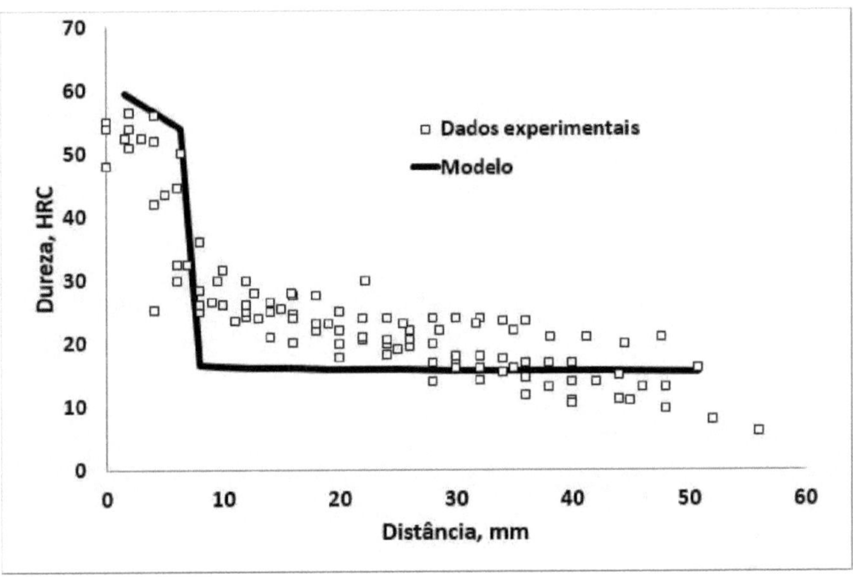

Source: Prepared by the author

Figure 34 - Transformed percentages of 1045 steel considering carbon redistribution

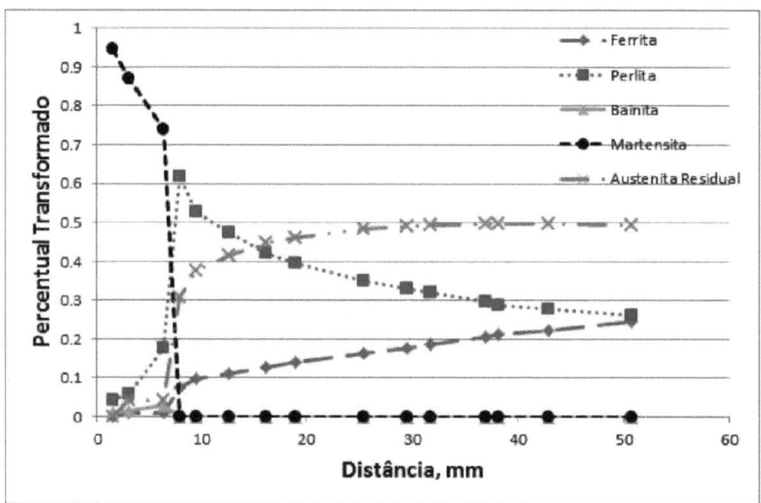

Source: Prepared by the author

In the same sequence, the results obtained for ABNT 4140 steel are presented. We note that the model follows the same hardness drop profile along the Jominy bar, but not at the same drop rate as in Figure 35.

Figure 35 - Jominy diagram 4140 steel

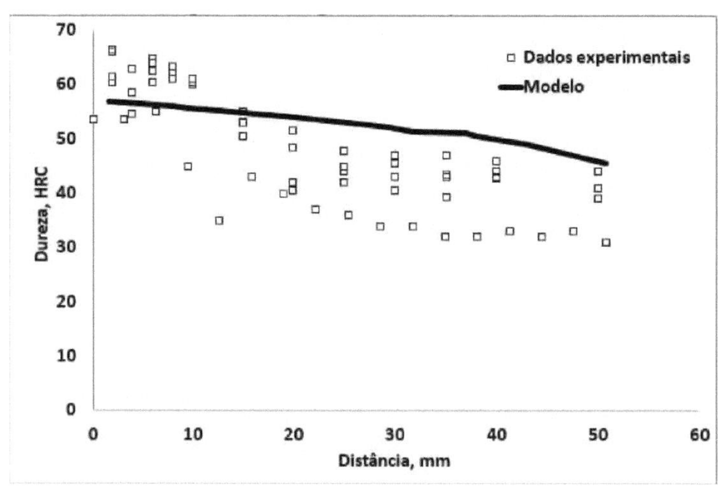

Source: Prepared by the author

Analyzing the evolution of the fractions of the Constituents generated along the length of the Jominy bar (Figure 36), it is possible to see a gradual reduction in the martensite fraction at the same time as there is an increase in the bainite content, causing the hardness to drop along the length.

Figure 36 - Transformed percentages 4140 steel

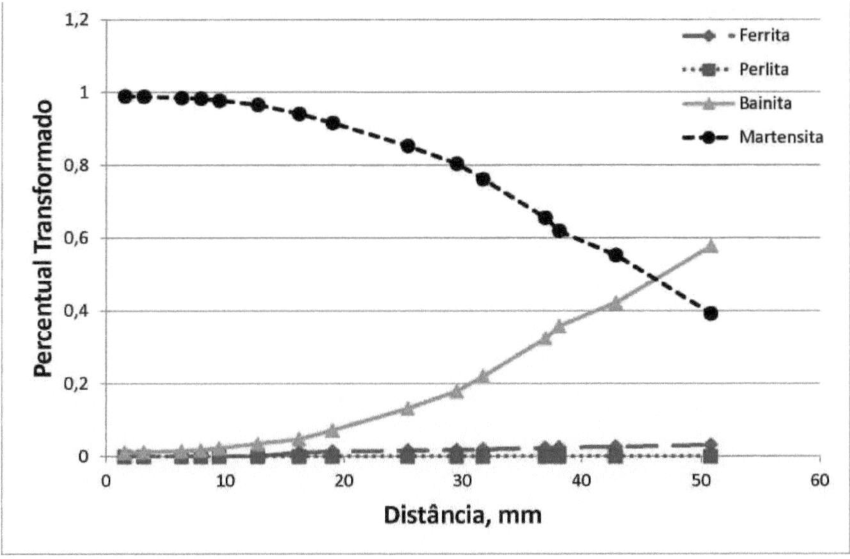

Source: Prepared by the author

After considering the redistribution of carbon, Figure 37, in the same way as ABNT 1045 steel, the change in the final result of ABNT 4140 steel was very subtle. Analyzing the variation in microstructures, Figure 38, we have a small amount of ferrite compared to the percentage amount of bainite and martensite, so the amount of residual austenite is also low.

Figure 37 - Jominy diagram 4140 steel considering carbon redistribution

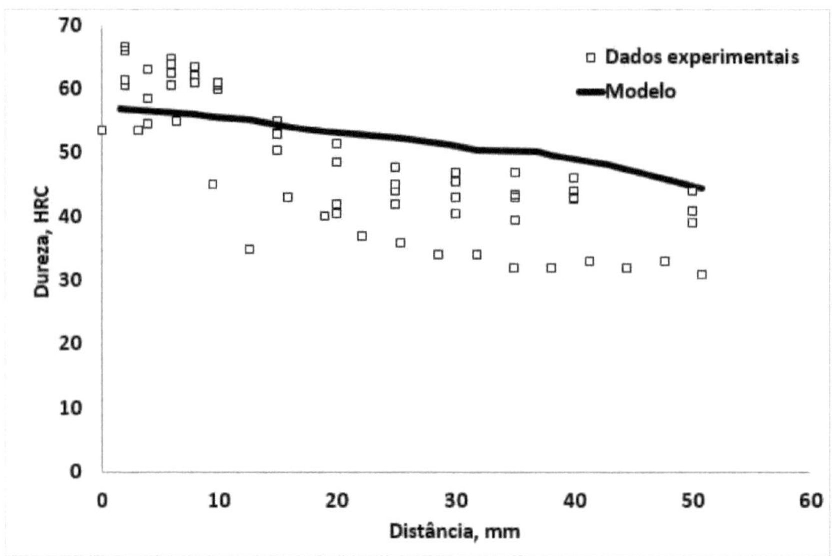

Source: Prepared by the author

Figure 38 - Transformed percentages of 4140 steel considering carbon redistribution

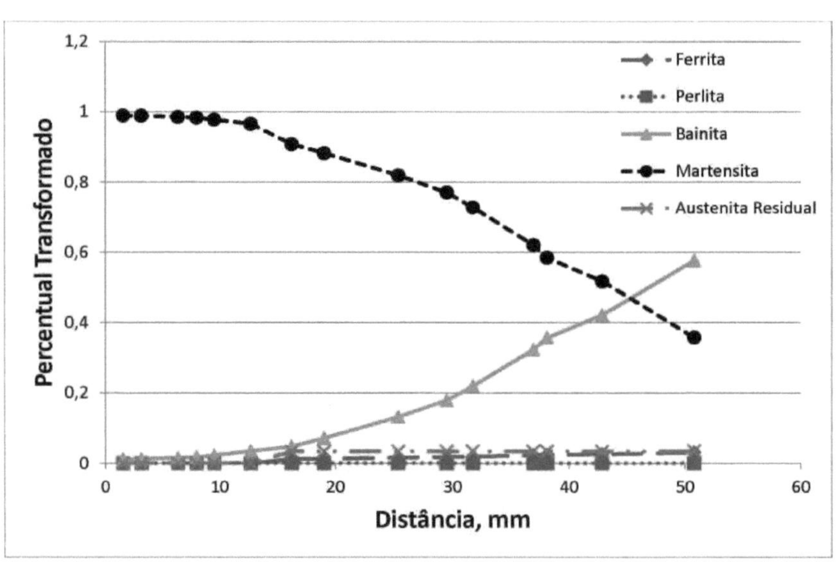

The AISI 4340 steel showed a slightly higher hardness in the model than the experimental values used for the comparison, as shown in Figure 39.

Figure 39 - Jominy diagram 4340 steel

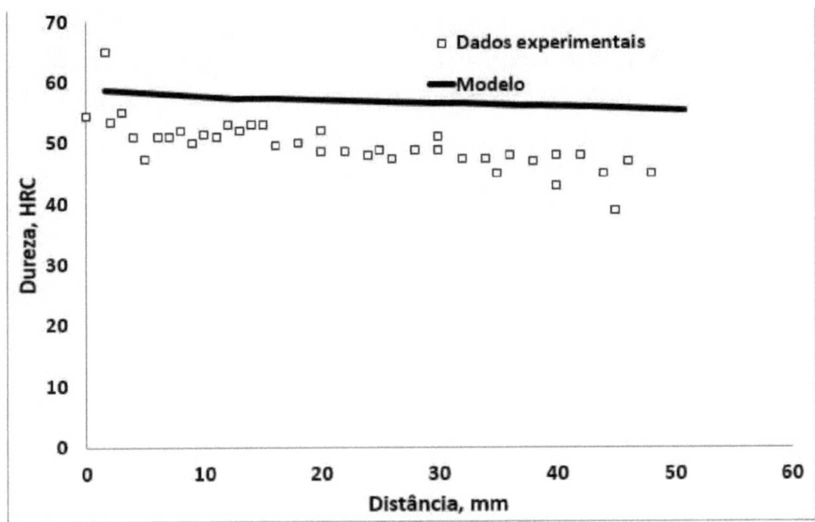

Source: Prepared by the author

Looking at Figure 40, you can see that the martensitic microstructure is predominant along the entire length and the other microstructures have values very close to zero.

Figure 40 - Transformed percentages 4340 steel

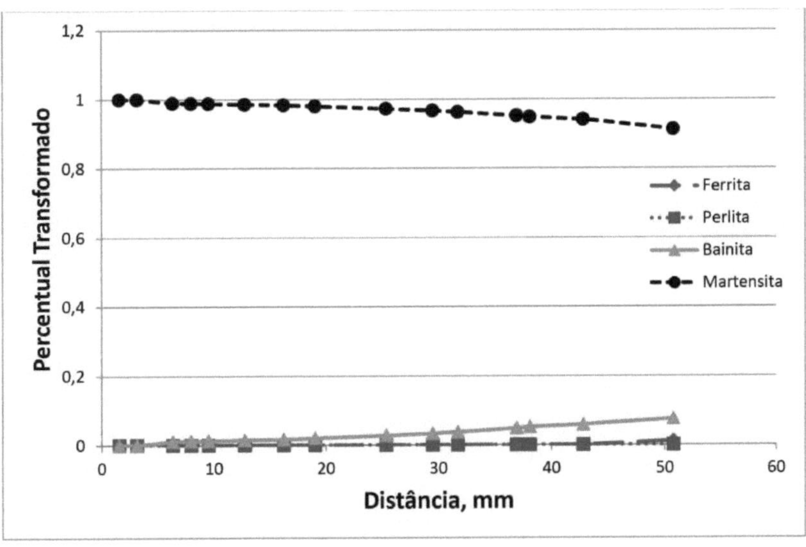

Source: Prepared by the author

Carrying out the same procedure as before, considering the redistribution of carbon, we again see that the change in the model was almost imperceptible (Figure 41). In the same way as with the previous steel, the residual austenite remained at values close to zero along with the other microstructures (Figure 42). We can therefore conclude that considering the redistribution of carbon for this model did not produce a considerable change in hardness in the final result.

Figure 41 - Jominy diagram for 4340 steel considering carbon redistribution

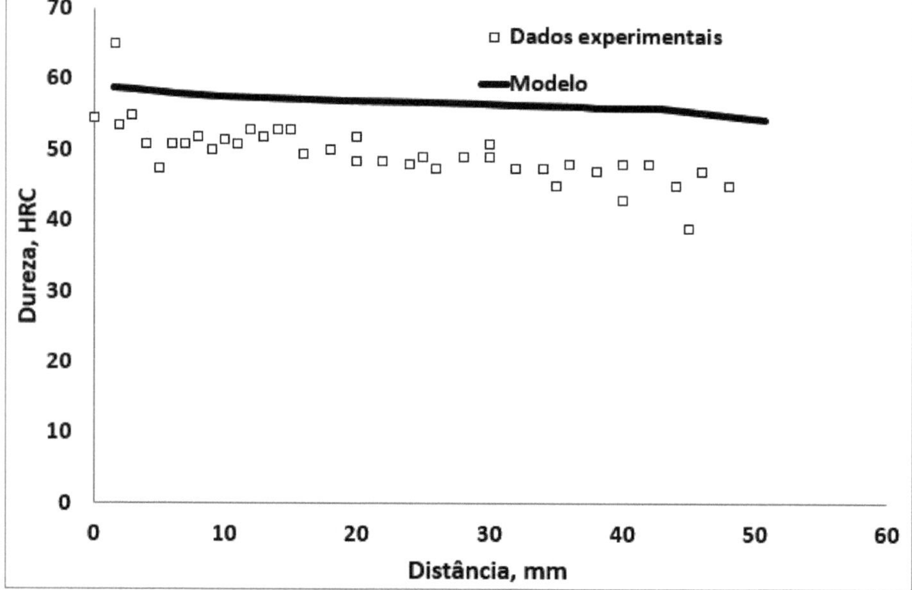

Source: Prepared by the author

Figure 42 - Transformed percentages of 4340 steel considering carbon redistribution

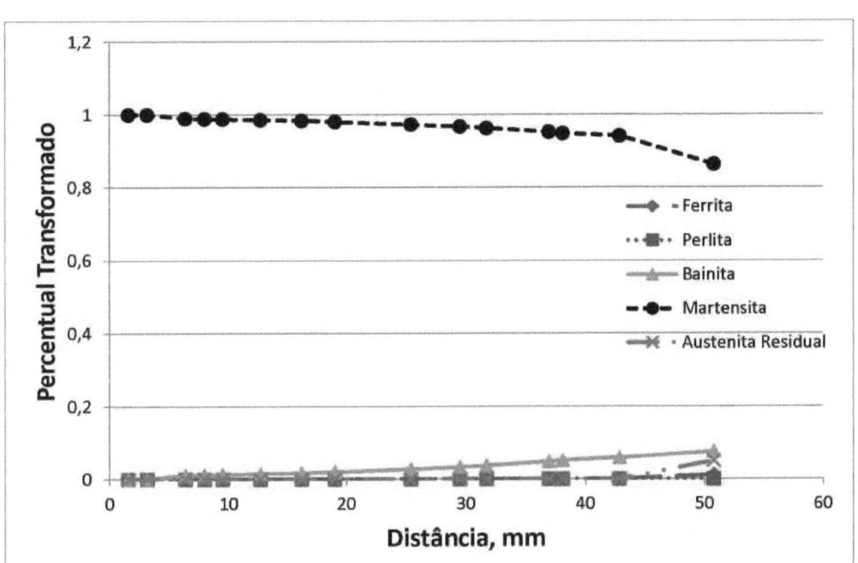

Source: Prepared by the author

5.3 Correction of TTT curves from Iemperability curves

The next steps to complete the work involved determining the transformed fraction of each phase or constituent of the steel as a function of the cooling rate and, consequently, calculating the expected hardness of the material. Once these calculations were implemented in the *software,* it was possible to obtain virtual Jominy hardenability curves, which can be used to validate the model developed. Validation can be carried out by directly comparing the predicted and experimental hardness values along the Jominy bar.

The *software* presents an optimization tool to adjust the curves of the virtual isothermal transformation diagram by comparing the result of a diagram generated from the Jominy test with the result of the Jominy diagram of the model generated from the intersection of the cooling curves with the isothermal transformation curves. An average absolute deviation parameter was used for the optimization algorithm, which represents how close or far the model is from representing the real values. The lower the value of this parameter, the better the representation of the model. In this way, the program's optimization function works by imputing gradual displacements to the curve selected by the user in order to reduce this parameter. On the data entry screen, the data for ABNT 1045 steel was entered using the hardness scale in Rockwell C and the length scale in millimeters and displaying the cooling curves.

Figure 43 - TTT curves 1045 steel

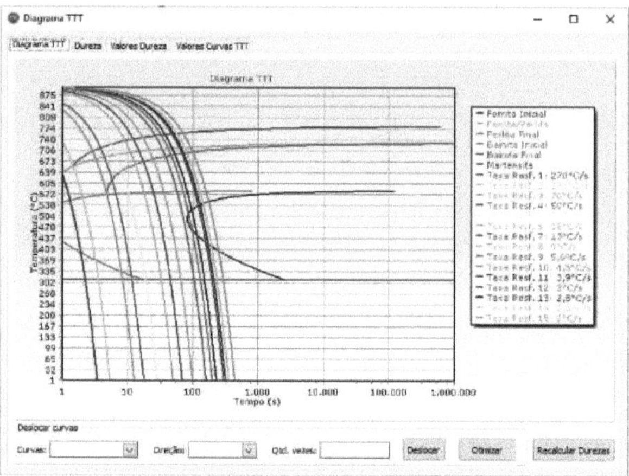

Source: Prepared by the author

After running the *software* calculations, the isothermal transformation curves and the Jominy hardenability curves are shown in Figure 43 and Figure 44.

Figure 44 - Jominy curve 1045 steel

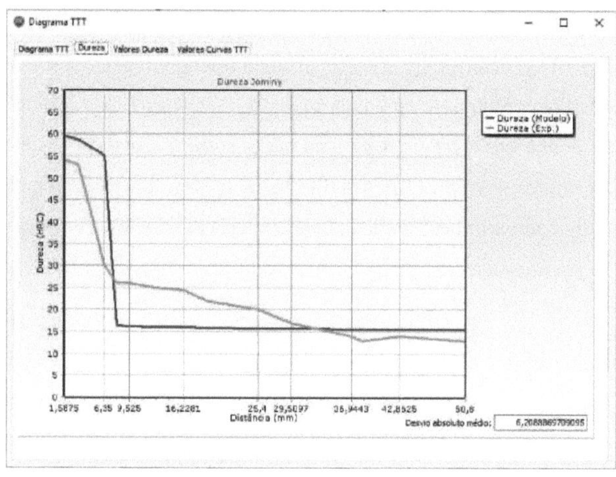

Source: Prepared by the author

Figure 44 shows the approximate value of 6.2089 for the mean absolute deviation parameter in the bottom right-hand corner. Reducing this value by shifting the TTT diagram curves would result in a better representation of the model. Comparing Figure 25 and Figure 43, we can see that in order to bring the model curves closer to the real curve, we need to shift the final bainite curve to the left.

Figure 45 shows that the optimization attempt was not effective, because even if we tried to shift this curve, there was no improvement in the average absolute deviation due to most of the cooling curves intersecting the final pearlite curve.

Figure 45 - TTT curves Optimized 1045 steel

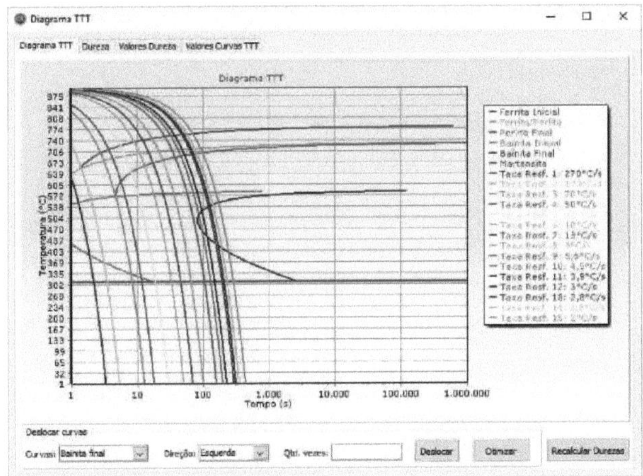

Source: Prepared by the author

Running the optimization again by shifting the ferrite curve to the left, it was possible to find an approximate mean absolute deviation value of 6.1907, which is slightly better than the previous value, Figure 46. However, by comparing the curves of the virtual TTT diagram with the experimental results, we can see that this shift is not ideal, see Figure 47.

Figure 46 - Optimized 1045 steel Jominy curve

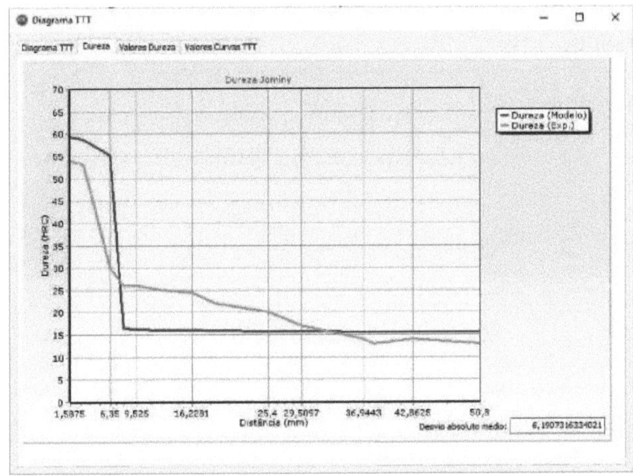

Source: Prepared by the author

Figure 47 - Modeled and real TTT diagram of optimized 1045 steel

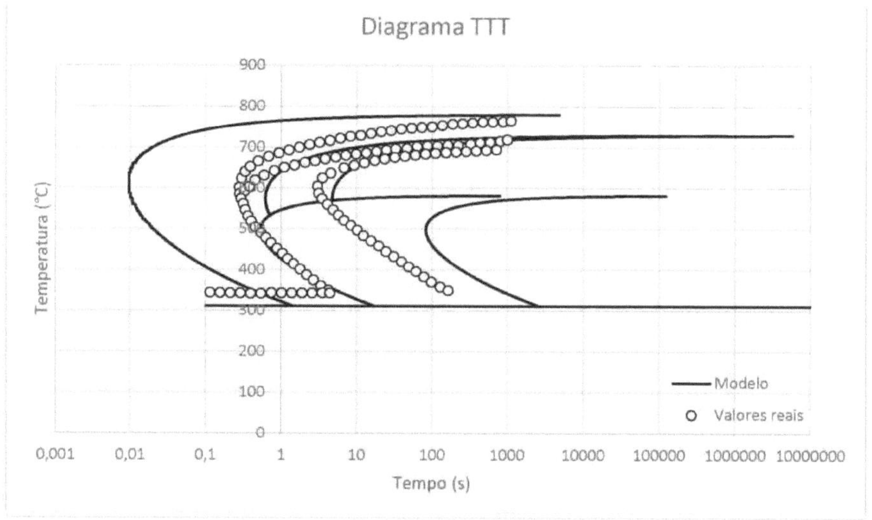

Source: Prepared by the author

Making a new attempt at optimization from the initial situation of the virtual curves, the Ferrite/Perlite curve was shifted to the left and a value of approximately 6.1554 in mean absolute deviation was found, as shown in Figure 48. Compared to the previous optimization, the result was better than the optimization of the Ferrite curve. Figure 49 shows the optimized TTT diagram of the model compared to the diagram based on real data and we can see that the virtual curve was more displaced than the initial diagram compared to the real diagram.

Figure 48 - Optimized 1045 steel Jominy curve

Source: Prepared by the author

Figure 49 - Modeled and real TTT diagram of optimized 1045 steel

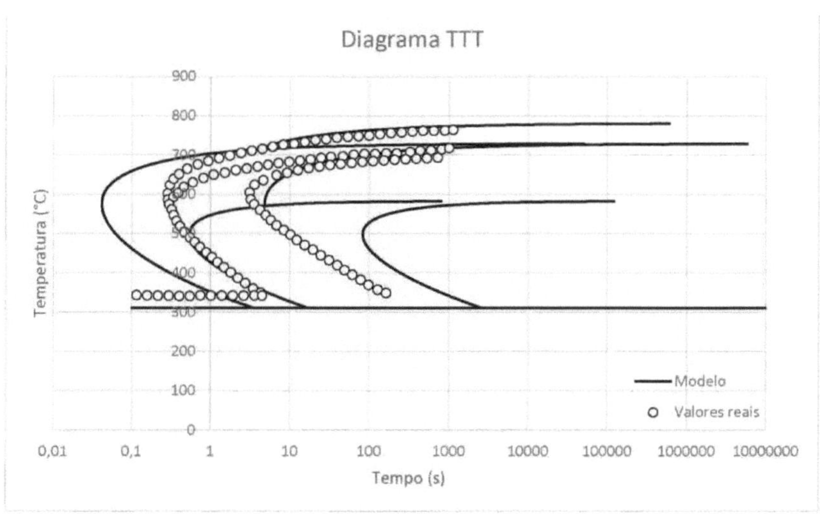

Source: Prepared by the author

Figure 50 below shows the Jominy graph with the initial average absolute deviation of approximately 5.5357 generated by the AISI 4140 steel model. After running the *software* optimization of the final bainite curve to the left, because we needed to reduce the hardness from the middle of the Jominy bar onwards, i.e. reduce the formation of martensite, a lower average absolute deviation value of approximately 5.4732 was obtained, so the final Jominy diagram obtained was as shown in Figure 51 and the final TTT curve compared to the real data was as shown in Figure 52.

Figure 50 - Jominy curve 4140 steel

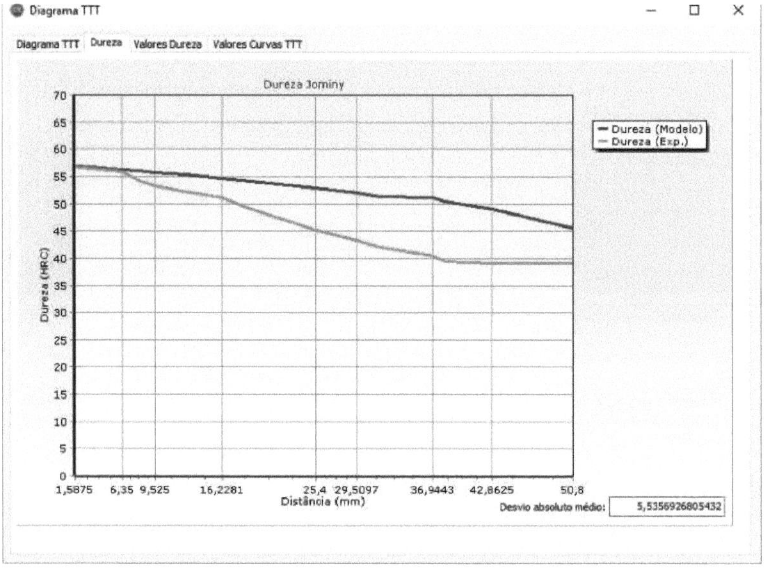

Source: Prepared by the author

Figure 51 - Jominy curve 4140 steel after optimization

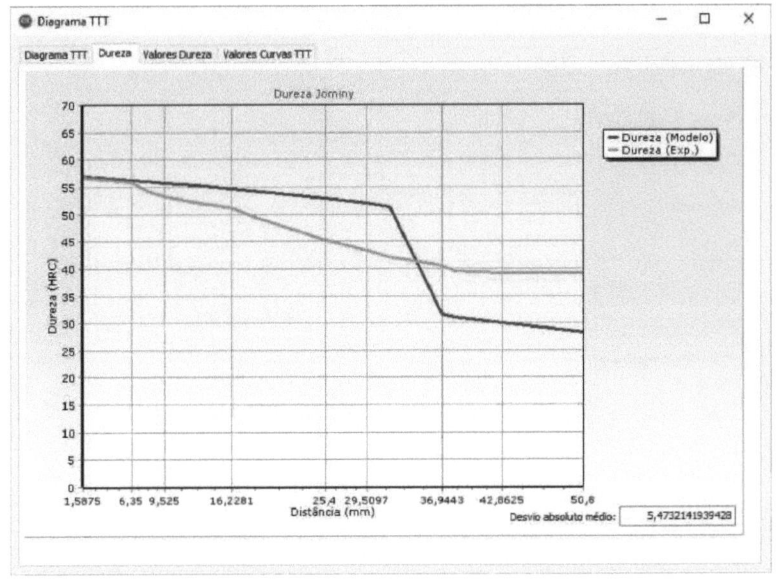

Source: Prepared by the author

Figure 52 - Optimized and real modeled TTT diagram for 4140 steel

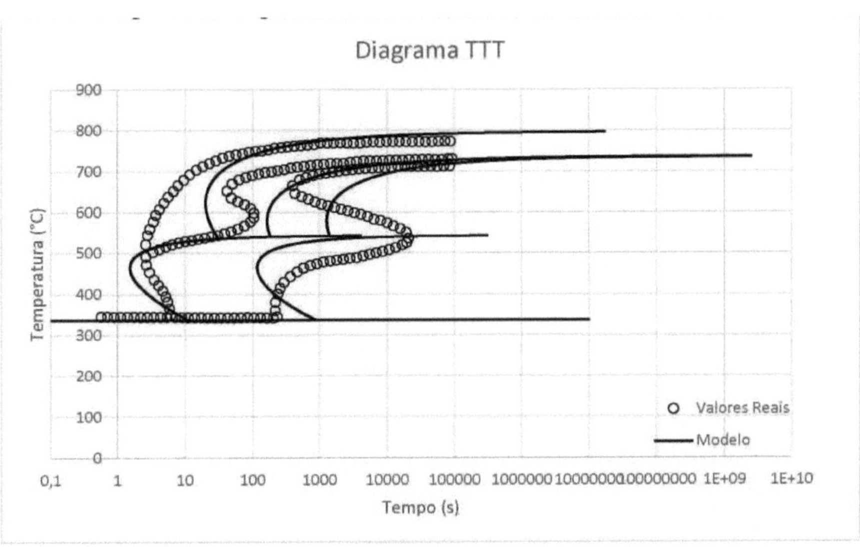

Source: Prepared by the author

For the AISI 4340 steel, the Jominy curve initially presented by the model is shown in Figure 53. Following the same optimizing path above, shifting the final bainite curve to the left, it was not possible to find an average absolute deviation value of approximately 2.9831 that is lower than that initially generated by the model.

Figure 53 - Jominy curve 4340 steel

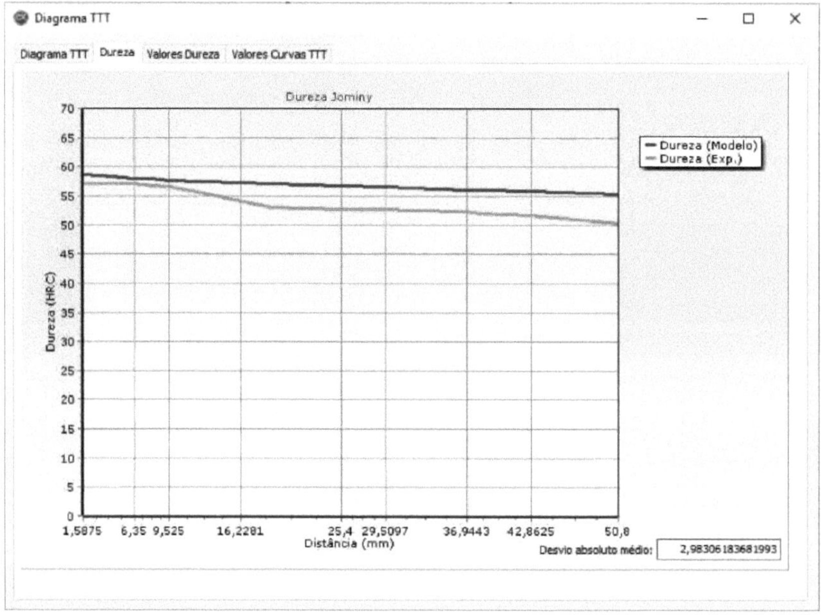

Source: Prepared by the author

6 CONCLUSION

Despite the difficulty in predicting TTT diagram curves from a mathematical model, the development of this work has made it possible to prove the effectiveness of the initial model developed by Victor Li (1998) for predicting virtual TTT curves, as well as predicting hardness for low-alloy steels. The benefits of this work extend as proposed, which is the estimation of the hardness of a steel from the microstructures generated according to each cooling rate of the material. The use of the algorithm developed in academic circles can contribute to understanding the issues involved in the heat treatment of low-alloy steels.

In general, the initial results obtained when running the *software* proved to be very accurate compared to the experimental results carried out in the laboratory for each of the steels used. In the case of the AISI 1045 steel, the optimization routine resulted in a slight improvement in the final hardness result compared to the real values and it was possible to see that the result that presents a virtual hardness closest to the experimental one will not always represent a virtual isothermal diagram closest to a real isothermal diagram. For the AISI 4140 steel, the values were also approximately as expected and, by running the optimization routine, it was possible to find an improvement in the representation of the results. However, there is a sudden reduction in hardness after using the optimization routine. This is due to the hardness calculation algorithm, as there is no gradual transition of martensite reduction, but rather a point where a martensitic microstructure value is considered and at a later point this value is equal to zero. Finally, AISI 4340 was the steel that showed the best approximation of the diagrams, TTT and Jominy hardness. And it was possible to verify that when trying to run the optimization routine, no better value was found.

The routine implemented in the *software* to take carbon redistribution into account proved to be inefficient, as the hardness values obtained were practically the same. These variations in hardness values, compared to the hardness curves without taking redistribution into account, were in the order of the third decimal place.

For future work, it would be interesting to develop a fully automatic optimization routine, unlike the one developed in this work, which requires a certain amount of user knowledge to define which curves it is necessary to shift in order to obtain a better result. Another implementation involving an improvement in the method for calculating the percentage of each microstructure generated from each cooling curve would mean that the hardness values would not vary abruptly depending on whether or not a particular microstructure was present.

Furthermore, once it has been possible to show that the transformation curves can be adjusted by comparing virtual hardenability curves with experimental hardenability curves, it is possible to seek to refine other parameters of the model. In this context, we highlight the possibility of refining the empirical parameters *PC, FC, BC* presented in equations 20, 23 and 26, computing the influence of chemical composition on the activation energy for carbon diffusion (Q = 27500 J/mol), among other factors.

BIBLIOGRAPHICAL REFERENCES

BANGARU, N. -R. V., SACHDEV, A. K., **Influence of cooling rate on the microstructure and retained austenite in an intercritically annealed vanadium containing HSLA steel**, Metallurgical and Materials Transactions A, v. 13, n. 11, p. 1899-1906, 1982.

BHADESHIA, H. K. D. H.; EDMONDS, D. V. **The Mechanism of Bainite Formation in Steels**. Acta Metallurgica, v. 28, p. 1265-1273, 1980.

BHADESHIA, H. K. D. H. **Lecutre 10: Overall Transformation Kinetics**. Available at: http://www.phase-trans.msm.cam.ac.uk/mphil/MP6-10.pdf. Accessed on: March 27, 2017.

BOK, H.-H. et al. **Non-isothermal kinetics model to predict accurate phase Transformation and hardness of 22MnB5 boron steel**. Materials Science & Engineering, v. 626, p. 67-73, 2015.

DE ANDRÉS, M. P.; CARSI, M. **Hardenability: an alternative to the use of grain size as calculation parameter**. Journal of Materials Science, v. 22, p. 2707-2716, 1987.

CALLISTER, William D. **Materials Science and Engineering. An Introduction.** 7. ed. John Wiley & Sons, Inc. 2007.

CARLONE, P.; PALAZZO, G. S.; PASQUINO, R. **Finite element analysis of the steel quenching process: Temperature field and solid-solid phase change**. Computer and Mathematics with Applications, v. 59, p. 585-594, 2010.

CHEN, Xiangjun et al. **The finite element analysis of austenite decomposition during continuous cooling in 22MnB5 steel**. Modelling and Simulation in Materials Science and Engineering, v. 22, p. 1-16, July 2014.

CHEN, Xiangjun et al. **A Modified Approach to Modeling of Diffusive Transformation Kinetics from Nonisothermal Data and Experimental Verification**. Metallurgical and Materials Transactions A, June 2016.

CHIAVERINI, Vicente. **Steels and Cast Irons: general characteristics, heat treatments, main types.** 7. ed. Brazilian Association of Metallurgy and Materials. 2008.

CLEMM, P. J.; FISHER, J. C. **The Influence of Grain Boundaries on the Nucleation of Secondary Phases**. ACTA Metallurgica, v. 3, p. 70-73, January 1955.

COLPAERT, Hubertus. **Metallography of common steel products**. 4. ed. Blucher, 2008.

DOMANSKI, T., BOKOTA, A., **Numerical models of hardening phenomena of tool steel based on the TTT and CCT diagrams**, Archives of metallurgy and materials, vol. 56, n. 2, p. 325-344, 2011.

GUO, Zhanli et al. **Modelling phase transformations and material properties critical to the prediction of distortion during the heat treatment of steels**. Int. J. Microstructure and Materials Properties, v. 4, n. 2, p. 187-195, 2009.

KUNG, C. Y., RAYMENT, J. J., **An Examination of the Validity of Existing Empirical Formulae for the Calculation of Ms Temperature**, Metallurgical Transactions A, vol. 13A, p. 328-331, 1982.

HAGENAH, H., MERKLEIN, M., LECHNER, M., SCHAUB, A., LUTZ, S., **Determination of the Mechanical Properties of Hot Stamped Parts from Numerical Simulations**, Procedia CIRP, v. 33, p. 167-172, 2015.

HAMELIN, C. J., MURANSKY, O., SMITH, M. C., HOLDEN, T. M., LUZIN, V., BENDEICH, P. J., EDWARDS, L. **Validation of a numerical model used to predict phase distribution and residual stress in ferritic steel weldments**, Acta Materialia, v. 75, n. 15, p. 1-19, August 2014.

HIPPCHEN, P., LIPP, A., GRASS, H., CRAIGHERO, H., FLEISCHER, M., MERKLEIN, M., **Modelling kinetics of phase transformation for the indirect hot stamping process to focus on car body parts with tailored properties**. Journal of Materials Processing Technology, v. 228, p. 59-67, February 2016.

LEE, Jye-Long Lee; BHADESHIA, H. K. D. H. **A methodology for the prediction of time-temperature-transformation diagrams**. Materials Science and Engineering, v. A171, p. 223-230, March 1993.

KIRKALDY, J. S.; BAGANIS, E. A. **Thermodynamic Prediction of the Ae3 Temperature of Steels with Additions o Mn, Si, Cr, Mo, Cu**. American Society for Metals and the Metallurgical Society of AIME, v. 9A, p. 495-501, April 1978.

KO, D.-H., KO, D.-C., KIM, B.-M., **Novel Method for Predicting Hardness Distribution of Hot-Stamped Part Using FE-Simulation Coupled with Quench Factor Analysis**. Metallurgical and Materials Transactions B, v. 46, n, 5, p. 2072-2083,

KOISTINEN, D. P.; MARBURGER, R. E. **A general equation prescribing the extent of the austenite-martensite transformation in pure iron-carbon alloys and plain carbon steels**. Acta Metallurgical, v. 7, p. 59-60, 1959.

KRAUSS, George. **Processing, Structure, and Performance.** ASM International. 2005.

MAZAR ATAKABI, M., YAZDIN, N., MA, J., KOVACEVIC, R., **High power laser welding of thick steel plates in a horizontal butt joint configuration**, Optics & Laser Technology, v. 83, p. 1-12, 2016.

NUNURA, César R. N.; SANTOS, Carlos A.; SPIM, Jaime A. **Numerical - Experimental**

correlation of microstructures, cooling rates and mechanical properties of AISI 1045 steel during the Jominy end-quench test. Materials and Design, v. 76, p. 230-243, 2015.

O'MEARA, N., ABDOVOLAND, H., FRANCIS, J. A., SMITH, S. D., WITHERS, P. J., **Quantifying the metallurgical response of a nuclear steel to welding thermal cycles**, Materials Science and Technology, v. 32, n. 14, p. 1517-1532, 2016.

PERELOMA, Elena and EDMONDS, David V. **Phase Transformations in Steels**. Woodhead Publishing Limited, 2012, v. 1, Chapter 4.

PORTER, D. A. and EASTERLING, K. E. **Phase Transformations in Metals and Alloys**. 2. ed. Boca Raton: CRC Press, 1992. 515p.

REED-HILL, Robert E. **Principios de Metalurgia Fisica**. 2. ed. Guanabara two S.A., 1982.

REED-HILL, Robert E.; ABBASCHIAN, Reza; ABBASCHIAN, Lara. **Physical Metallurgy Principles**. 4. ed. Cengage Learning, 2008.

SERAJZADEH, S.; TAHERI, A. Karimi. **A study on austenite decomposition during continuous cooling of a low carbon steel**. Materials and Design, v. 25, p. 673-679, 2004.

SUSHANTHI, Neethi; MAITY, Joydeep. **A Finite Difference-Based Modeling Approach for Prediction of Steel Hardenability**. Journal of Materials Engineering and Performance, v. 23(6), p. 2209-2218, June 2014.

SUSHANTHI, N., MAITY, J. **An independent modeling approach for prediction of hardenability in steels**. Steel Research International v. 86, n. 4, p. 329-340, April 2015.

TRZASKA, J.; DOBRZANSKI, L.A. **Modelling of CCT diagrams for engineering and constructional steels**. Journal of Materials Processing Technology, v. 192-193, p. 504-510, 2007.

TRZASKA, J. **Calculation of the steel hardness after continuous cooling**. Archives of Materials Science and Engineering, v. 61, p. 87-92, June 2013.

UMEMOTO, Minoru; NISHIOKA, Nobuo; TAMURA, Imao. **Prediction of Hardenability from Isothermal Transformation Diagrams**. J. Heat Treating, v. 2, n. 2, p. 130-138, December 1981.

VICTOR LI, M. et al. **A Computational Model for the Prediction of Steel Hardenability**.

Metallurgical and Materials Transactions, v. 29B, p. 661-672, June 1998.

WANG, Q., LIU, X. S., WANG, P., XIONG, X., FANG, H. Y., **Numerical simulation of residual stress in 10Ni5CrMoV steel weldments**, Journal of Materials Processing and Technology, v. 240, p. 77-86, 2017.

Printed by Books on Demand GmbH, Norderstedt / Germany